THE MEDICINE CABINET

Contributors:

Understanding Our Bodies: Contributions from Muriel Bailly, Katy Barrett, Robert Bud, Rupert Cole, Katie Dabin, Jannicke Langfeldt, Isabelle Lawrence, Sarah Wade

Birth and Death: Contributions from Muriel Bailly, Imogen Clarke, Katie Dabin, Stewart Emmens, Selina Hurley, Natasha McEnroe, Isabelle Lawrence

Diagnosis: Contributions from Muriel Bailly, Tim Boon, Katie Dabin, Natasha McEnroe, Emma Stirling-Middleton

Surgery: Contributions from Katy Barrett, Jessica Bradford, Katie Dabin, Stewart Emmens, Selina Hurley

Public health: Contributions from Tim Boon, Imogen Clarke, Stewart Emmens, Natasha McEnroe, Sara Öberg Strådal, Sarah Wade

Assistive technology: Contributions from Gemma Almond, Sarah Bond, Jack Davies, Stewart Emmens, Selina Hurley, Rebecca Kearney, Isabelle Lawrence

Belief: Contributions from Sarah Bond, Katie Dabin, Stewart Emmens, Selina Hurley, Isabelle Lawrence, Sara Öberg Strådal, Annie Thwaite

Drugs and Pharmacy: Contributions from Muriel Bailly, Robert Bud, Jessica Bradford, Imogen Clarke, Jack Davies, Selina Hurley, Isabelle Lawrence, Emma Stirling-Middleton, Natasha McEnroe, Sarah Wade

War: Contributions from Jack Davies, Stewart Emmens, Selina Hurley, Natasha McEnroe

Hospitals: Contributions from Sarah Bond, Imogen Clarke, Katie Dabin, Stewart Emmens, Selina Hurley, Natasha McEnroe

THIS IS AN ANDRE DEUTSCH BOOK

Published in 2019 by André Deutsch
An imprint of the Carlton Publishing Group
20 Mortimer Street
London W1T 3JW

A CIP catalogue for this book is available from the British Library.

ISBN 978 0 233 00610 9

Printed in Italy

SCIENCE
MUSEUM

THE MEDICINE CABINET

The Story of Health and Disease Told Through Extraordinary Objects

Edited by
Natasha McEnroe
and Selina Hurley

ANDRE
DEUTSCH

CONTENTS

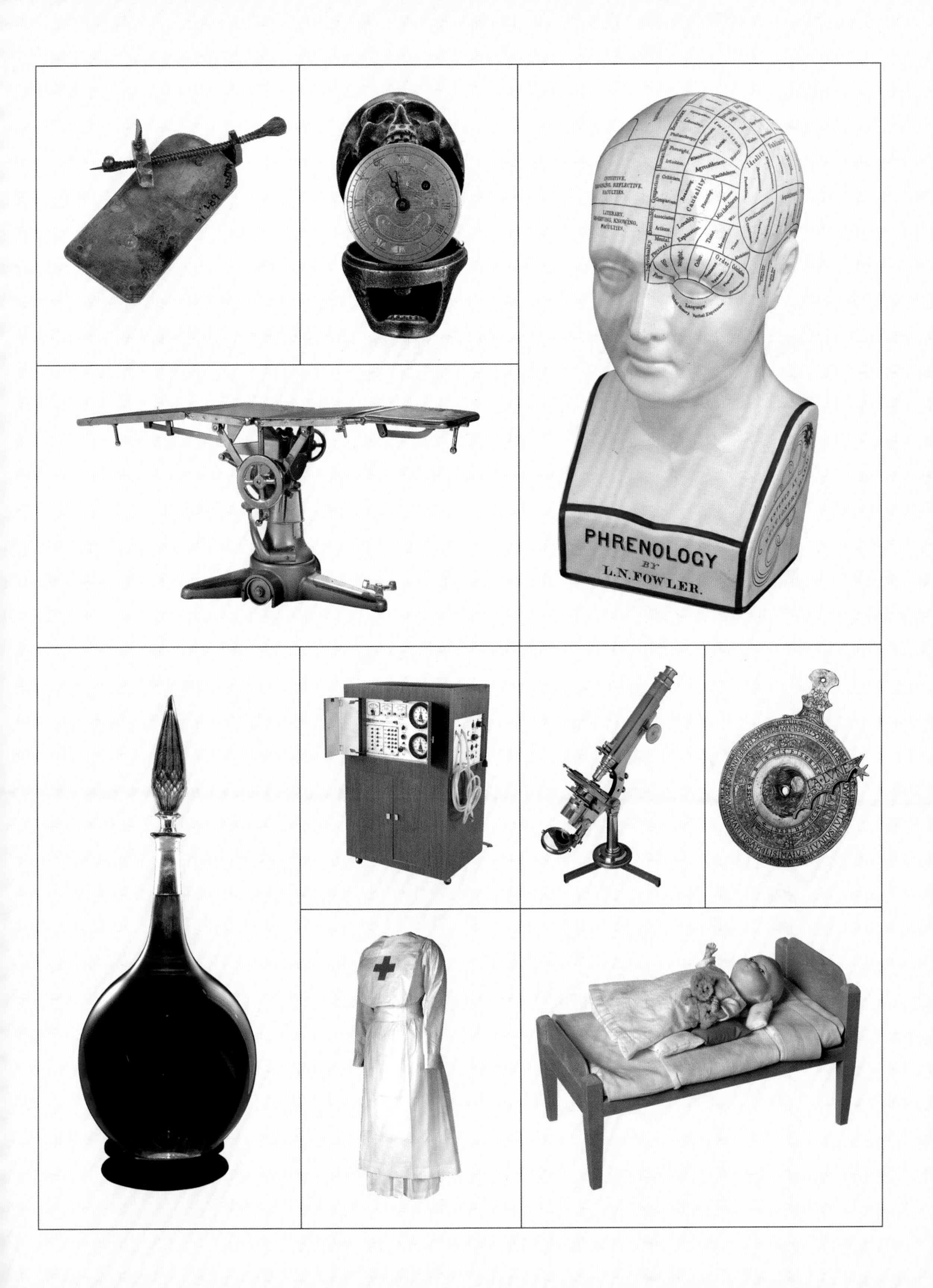
PHRENOLOGY
BY
L.N. FOWLER.

INTRODUCTION

Many renowned museums have their origins in cabinets of curiosities. These were eclectic mixtures of carefully displayed items celebrating the natural world, art, ethnography, religion and culture. The medicine collection of the Science Museum, London, is just as wide-ranging, and covers many of these same topics as well as the more expected surgical instruments and pharmacy-ware. Much of the variety found in this collection and in this book is owed to Henry Wellcome. Wellcome was a pharmaceutical entrepreneur, investing much of his profit in his attempt to form a collection that reflected the whole human experience of life and death, health and illness, religion and science. While his ambition was beyond the limits of any one person or even any one institution, his legacy is the largest and most intriguing collection of artefacts relating to the history of medicine. After his death this collection, which totalled over a million objects, was dispersed to museums and private collections across the world. The core part of the collection was transferred to the Science Museum on long-term loan in 1979, heralding a new era for the history of medicine at the museum. Since then, curators have added to the collection, which now boasts items such as the world's first MRI scanner, a travelling X-ray van, molecular models of haemoglobin and penicillin, the first robotic surgical assistant used in the UK and one of the most internationally important collection of prosthetic limbs.

The 150,000 medical objects in the Science Museum's care document humanity's ongoing concern with our health. In this book, 21 authors associated with the Science Museum share over 100 stories from the collection. Through these stories, we can see how people have attempted to solve some of our biggest medical challenges: to make sense of our bodies, to make a diagnosis, to care for us at all phases of our lives both as individuals and as communities, and to develop an ever-widening range of treatments.

Depending on where we are in the world, where we are in our own lives, and where we find ourselves in history, our expectations of medicine continue to change. There are times when medicine has met, fallen short or surpassed our expectations. It has and continues to change how we understand our sense of self and our bodies, and how we relate to others, particularly those who practise medicine. Each of the stories in the following pages tell us as much about the history of medicine as the history of ourselves.

Above: Post-First World War artificial arm, in the Science Museum's store.

Opposite top: Medical glassware held in the Science Museum's collection.

Opposite below: Surgical instruments for controlling blood flow, held in the Science Museum's collection.

VOICE
JUBES.
AROMATIC
PASTILLES
ZINGIB
SYRUPUS
SYR AURA

A 614107
A 614106
A614094
A614093

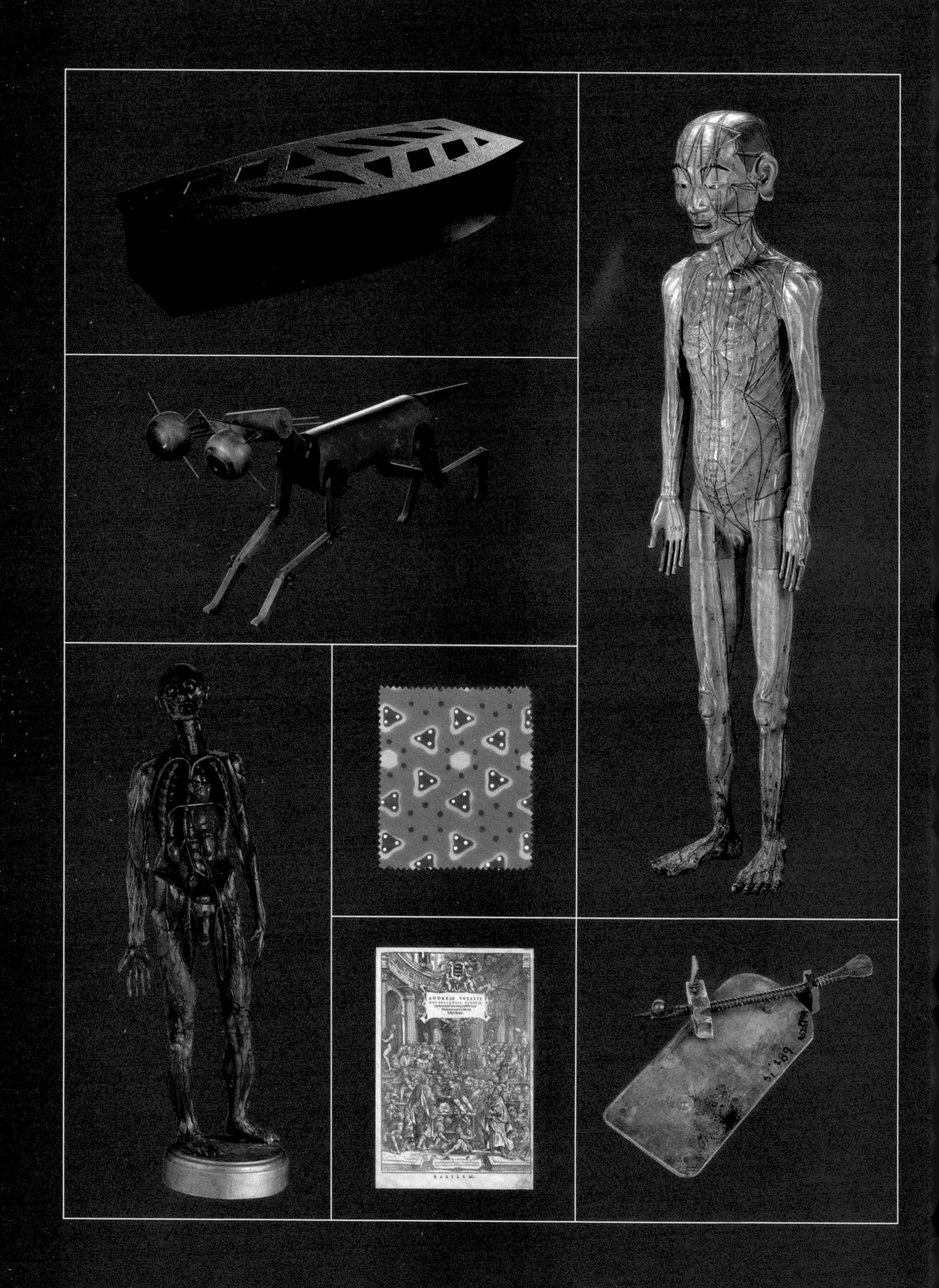
ANDREAE VESALII
BASILEAE

1

UNDERSTANDING OUR BODIES

The wonder of the human body, how it works and why it does not work, has been a source of fascination for amateurs and professionals alike for centuries. Examining bodies, living or dead, animal and human, has led to the greater understanding of life and health. Looking at, smelling and even tasting bodily fluids has long been the means of monitoring and testing sickness and health. Beautiful and strange anatomical models, made of wax or papier mâché, chart our understanding of systems of nerves, muscles and blood. Recording and measuring physical features like height, strength and growth have led to greater understanding of the health of the wider population. Technology such as X-ray and MRI scans enable us to see beneath the skin of the body, revealing the hidden structures within.

AUZOUX'S ANATOMICAL MODEL

Louis Auzoux produced beautiful anatomical models from papier mâché to teach his medical students about the human body.

In order to treat their patients, doctors need to know how the body functions both inside and out. One of the most straightforward ways of studying human anatomy is to examine the body after death, by cutting it open in a process known as dissection. Until the 1500s, this practice was condemned by religious belief and faced great public hostility. Although the study of anatomy later became more acceptable, the challenge of keeping bodies fresh long enough to be used for teaching remained. Hardened to the smell of decay and seeping bodily fluids, anatomists had to work fast to dissect the body before it became too decomposed to use in lectures and demonstrations.

To overcome this problem, Louis Auzoux, a young French physician, devised detailed anatomical models constructed from papier mâché to support the study of anatomy. The models comprised of multiple detachable parts, each clearly labelled and sturdy enough to withstand repeated use. In 1825, after years of practising, Auzoux presented his first model of a full human figure to the French Académie Royale, along with other works. The Académie declared Auzoux's efforts commendable but commented that the aesthetic of the models could be improved.

From this encouraging start, Auzoux continued to produce papier mâché models, diversifying his practice to include botanical and zoological specimens. In 1828 he set up a factory in his hometown of Saint-Aubin-d'Ecrosville and soon employed more than 100 staff. He hired people with no medical or craftsmanship skills and trained them on the job, arguing that his workers learnt about anatomy from making the models. He claimed that his models could replace "experts" and that people could learn directly from manipulating them. Auzoux produced the first edition of each new model himself. Moulds were then taken from his original so that models could be produced in series and disseminated around the world. This example is one of Auzoux's prototypes, from which all models were produced.

The worldwide dissemination of his models brought Auzoux international fame. His success at the 1851 Great Exhibition in London, which won him much acclaim and many prizes, shows how experts and laypeople alike were eager to see his work. Although his creations were originally intended for medical training, he subsequently marketed them to a general audience, claiming that even beginners could learn anatomy and physiology without the help of a teacher.

Left: Postcard showing the Auzoux factory in Saint-Aubin-d'Ecrosville, 1870–1900.

Opposite: Anatomical model of a male figure by Louis Auzoux, 1825–60.

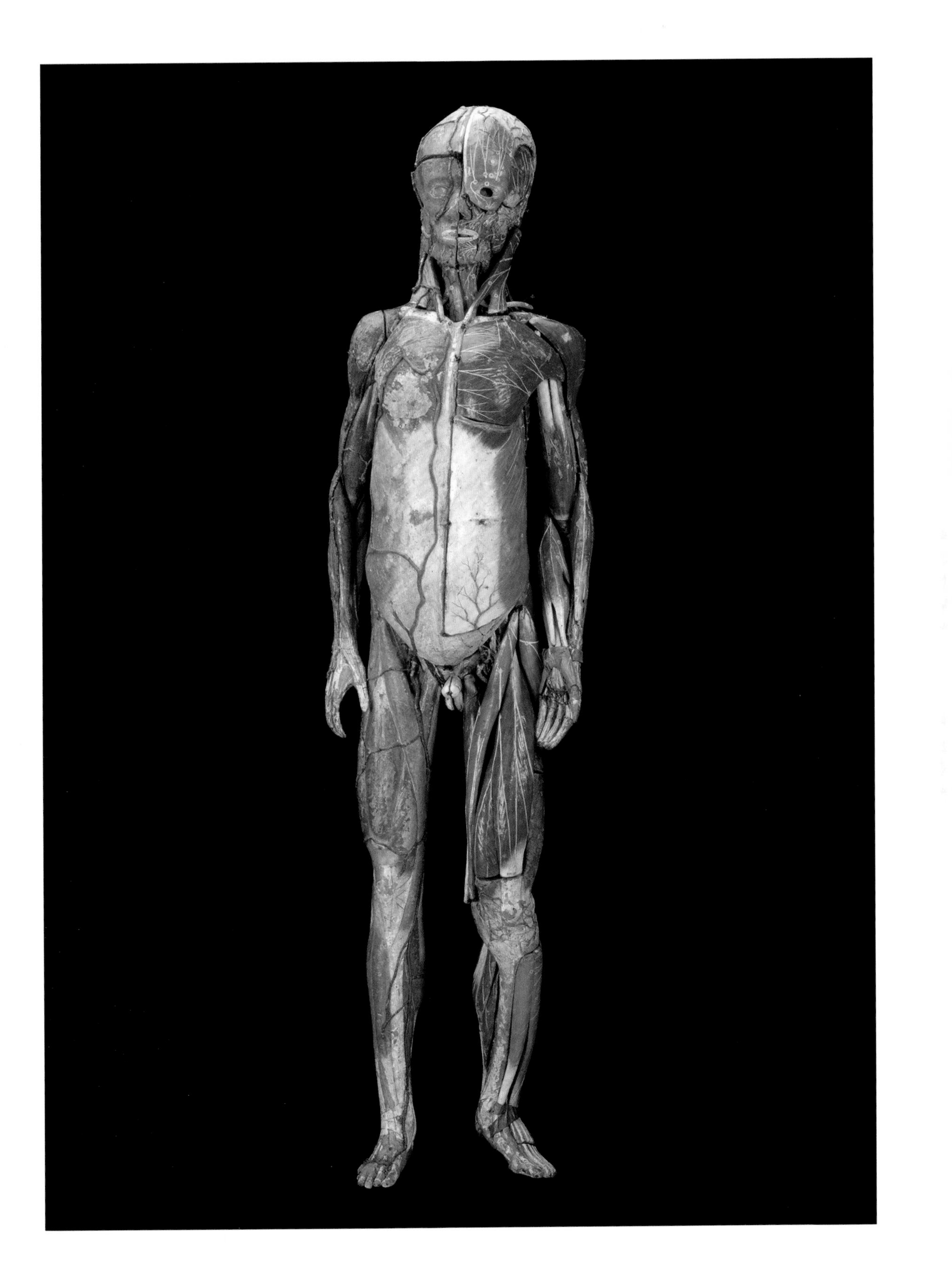

ANDREAE VESALII
BRVXELLENSIS, SCHOLAE
medicorum Patauinæ professoris, de
Humani corporis fabrica
Libri septem.
CVM CAESAREAE
Maiest. Galliarum Regis, ac Senatus Veneti gra
tia & priuilegio, ut in diplomatis eorundem continetur.
BASILEAE.

VESALIUS'S *ON THE FABRIC OF THE HUMAN BODY*

Andreas Vesalius's volume of anatomy is a landmark book that explained and illustrated the workings of the human body.

This book, *De Humani Corporis Fabrica*, represents the birth of modern anatomy as well as the first comprehensive visual account of the human body. Andries van Wesel (Latinized to Vesalius) was Professor of Anatomy at the University of Padua in Italy and a passionate advocate for dissection as the foundation of medical knowledge. He sought to revive the practices and knowledge of ancient Greece and Rome but also to update them, crucially challenging ideas of the body that had been established by Galen – a Greek physician, surgeon and philosopher – in 200 CE, based on dissecting animals. Vesalius used his publication *De Humani Corporis Fabrica* as a manifesto for his ideas. The use of extraordinarily detailed printed images was crucial to the work's success.

The frontispiece provides a powerful visual presentation of Vesalius's argument. At the centre of a crowded anatomical theatre, the author dissects a female cadaver. His right hand points to the dissected body, his left to heaven, making clear that knowledge of the body leads to knowledge of God's creation. Surrounding figures take notes and compare his demonstration to printed texts. Dressed in togas, the ancient authorities Galen, Aristotle and Hippocrates look on approvingly. Under the table, two barber-surgeons vie to sharpen the anatomist's knives. Vesalius wanted a critical shift in the structure of medical instruction, in which the anatomical professor would carry out dissection, rather than directing and commenting on the work of a surgeon.

The *Fabrica* included 82 further woodblock images, as well as hundreds of illuminated initials. The artist is unknown, but suggestions have included Jan Stephen van Calcar, pupil of Titian, and the landscape artist Domenico Campagnola. It is likely that multiple artists contributed, including Vesalius himself. He certainly supervised and paid for the woodblocks, providing unusual control over the form and placement of his images. He created enduring images of a canonical human body, with detailed dissections of the skeleton, muscles, viscera, circulatory, reproductive and nervous systems, and of specific organs. A cut-out broadsheet allowed organs to be pasted on to a single figure, creating a paper version of the dissection experience that Vesalius advocated. Figures appeared in picturesque landscapes or as memento mori, playing on Christian imagery concerned with the frailty of human ambition. They were beautifully-crafted images designed to appeal to wealthy scholars and physicians, and detailed tools for supporting Vesalius's controversial medical arguments. These images established conventions that continued to influence artists for centuries.

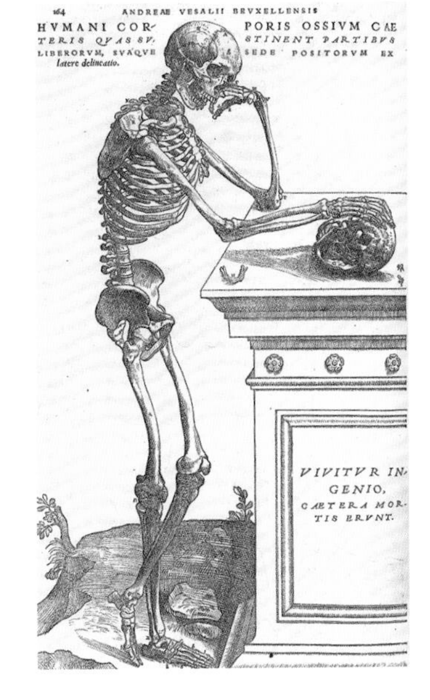

Opposite: Frontispiece from *De Humani Corporis Fabrica* [On the Fabric of the Human Body], by Andreas Vesalius, 1543, folio.

Right: Contemplative skeleton, illustration on p.164 of *De Humani Corporis Fabrica* [On the Fabric of the Human Body], by Andreas Vesalius, 1543, folio.

PASTEUR'S MICROSCOPE?

Louis Pasteur made incredible breakthroughs in medicine, and his scientific equipment was treasured by his admirers.

Hero-worship of a sports person or celebrity is nothing new, but revering a scientist was a new phenomenon in the nineteenth century. Louis Pasteur, the French chemist and microbiologist, had a remarkable career. After discovering that microbes were responsible for souring alcohol, inventing the technique of pasteurization to sterilize milk, developing vaccines for rabies and anthrax, and saving the silkworm industry from disease, Pasteur's scientific reach had made him one of the most famous people, let alone scientists, on the planet by the time of his death.

When someone is that celebrated, it is not surprising that all manner of materials and relics associated with them are preserved to record their greatness. This is a microscope made by Nachet & Son, a French scientific instrument company, which was claimed to have been used by Pasteur on his work on microbes and investigations into diseases of silkworms. But a closer analysis of its origins tells a different story.

The microscope was purchased by Henry Wellcome, an American-British chemist and entrepreneur who founded the pharmaceutical company Burroughs, Wellcome & Co. Wellcome spent millions buying objects that documented the history of medicine to create his Wellcome Historical Medical Museum, which opened in 1913. Appreciating Pasteur's fame as a great man of modern medicine, Wellcome sent his suave agent Captain Johnston-Saint to source Pasteur souvenirs to add to his collection.

When Pasteur died, his associates preserved any equipment associated with him. Silkworm specimens, glassware, apparatus and microscopes used in everyday laboratory life were kept as evidence of his greatness. Many were given to museums to preserve Pasteur's work for posterity. Others cashed in on the market for Pasteur relics. The owner of Nachet & Son offered to sell Wellcome microscopes reportedly used by Pasteur.

Over a succession of visits and dinners Johnston-Saint and Nachet did business, but it became increasingly unclear whether Nachet was offering originals or replicas. Whether Nachet knowingly and unscrupulously passed on this microscope, or whether direct personal association did not matter to Captain Johnston-Saint, is unclear. What is clear is that it is unlikely that this particular microscope was used by Pasteur and certainly not for claimed experiments on spontaneous generation; more probably it was an instrument used by silkworm farmers to assess the presence of silkworm disease according to Pasteur's method.

Opposite: Louis Pasteur, with two dogs (referring to his work on rabies), a palm and a snake around a bowl (indicating achievement in hygiene), 1890s.

Right: Compound monocular microscope made by Nachet & Son, 1861–70.

MORTSAFE

To protect the recently dead from unscrupulous grave-robbers, families would use heavy cages in which to contain the coffin.

Made of iron and large enough to contain a coffin, this cage was known as a mortsafe. Grieving relatives in the 1800s used devices like this to protect the bodies of their loved ones from being stolen and sold to medical professionals for the purpose of dissection. Coffins were kept inside the cage for a few days until the corpse began to decompose beyond the point of usefulness for anatomical study, after which time they were buried. Mortsafes reflected a very real fear in the 1800s about what might happen to one's body after death.

The growth in body-snatching was a response to a remarkable transformation in medical training during the eighteenth century. For hundreds of years, doctors learned human anatomy from information held in books and manuscripts. By the 1700s, the situation had profoundly changed. Medical students were expected to glean their knowledge from first-hand experience of dissecting corpses rather than relying on textbooks. By 1828 there were around 800 students studying medicine in London alone. Medical students paid good money for hands-on training, leading to a plethora of new schools where dissection was taught opening all over Europe. These anatomy schools relied on a regular supply of fresh corpses for students to practise on. At the start of the nineteenth century, however, the only bodies legally available for dissection were those of recently executed criminals. Yet there were too few hangings to meet the growing demand. Fighting and even rioting erupted at the scaffold, as family and friends of the deceased tried to prevent unscrupulous men taking corpses to sell to the medical schools.

The lucrative market created by anatomy schools gave rise to a new business opportunity for so-called resurrectionists, who illegally plundered fresh graves to sell the bodies within. Complicit in this crime, leading physicians argued that it was better to use corpses for dissection – however obtained – than for the living to suffer the consequences of ignorant doctors. However, some grave-robbers went further, resorting to murder to fulfil orders. Fierce public outrage forced the Government to take action. The 1832 Anatomy Act licensed the use of unclaimed bodies of the poor, who had died in workhouses or charitable hospitals, for dissection. Flooding the market with corpses, the Act proved to be the final nail in the coffin for grave robbing.

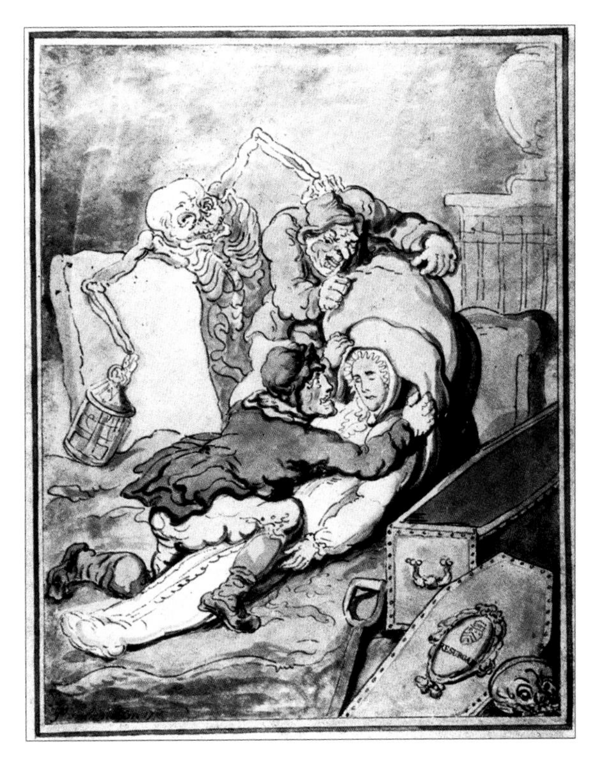

Opposite: Mortsafe, 1800–22.

Above: Death, as a nightwatchman holding a lantern, grabs one of the grave-robbers from behind, T. Rowlandson, 1775.

MOLECULAR WALLCOVERING AT THE FESTIVAL OF BRITAIN

Intricate patterns of molecules of the human body became a design smash hit in 1950s Britain.

Briefly, and perhaps unexpectedly, insulin became an icon of 1950s interior design. From restaurant walls to cinema seats, insulin's molecular structure inspired several patterns used in the decoration for the 1951 Festival of Britain.

Described as a tonic to the nation after post-war austerity, the Festival of Britain was a giant exhibition celebrating Britain's past, present and future, and emphasizing national achievement in science, technology, industry and the arts. The Festival took place across the UK, with its central exhibition located at London's South Bank. Among the many organizing committees was the Festival Pattern Group, conceived as a collaborative enterprise between scientists, manufacturers and industrial designers to create a "look" for the Festival based on structural diagrams of molecules. The Group's scientific consultant was Helen Megaw.

Megaw was a specialist in X-ray crystallography, a scientific field where molecules in crystal form are bombarded with X-rays to reveal their inner arrangement of atoms on photographic images. Long before the Festival, Megaw and her fellow crystallographers were aware of the artistic and design potential of the highly symmetrical atomic diagrams they produced from these X-ray photographs. Through the Festival Pattern Group she liaised with many of the leading lights of crystallography, including Dorothy Crowfoot Hodgkin. The two had worked together at Cambridge in the 1930s. When Crowfoot Hodgkin married in 1937, Megaw's wedding gift was a linen cushion embroidered with the crystal structure of aluminium hydroxide.

Crowfoot Hodgkin spent the best part of her career working on insulin, the hormone that helps the body absorb glucose. It was first discovered by a research team in 1921. Crowfoot Hodgkin took her first X-ray photograph of insulin crystals in 1934. It would take her another 35 years of painstaking analysis before she finally determined its three-dimensional structure. She was awarded the Nobel Prize in Chemistry in 1964 for her work in solving the structure of two other molecules that became medically important: penicillin and vitamin B12. But insulin perhaps remains her greatest achievement, given its atomic complexity. In her later life she travelled the world giving talks on its importance for diabetes.

Above: Dorothy Crowfoot Hodgkin in her Oxford laboratory, 1965.

Opposite: Sample of Rexine wallcovering, made by Imperial Chemical Industries, in "Insulin 8.25" design for the Festival of Britain, 1951.

This particular fabric sample adorned the Science Museum's walls during the 1951 Festival. It was made by ICI from their highly flammable (and now known to be toxic) "Rexine" artificial leathercloth. Megaw told Crowfoot Hodgkin that the Festival Pattern Group manufacturers were "very much taken by insulin – except for its name – which they thought ... might put people off".

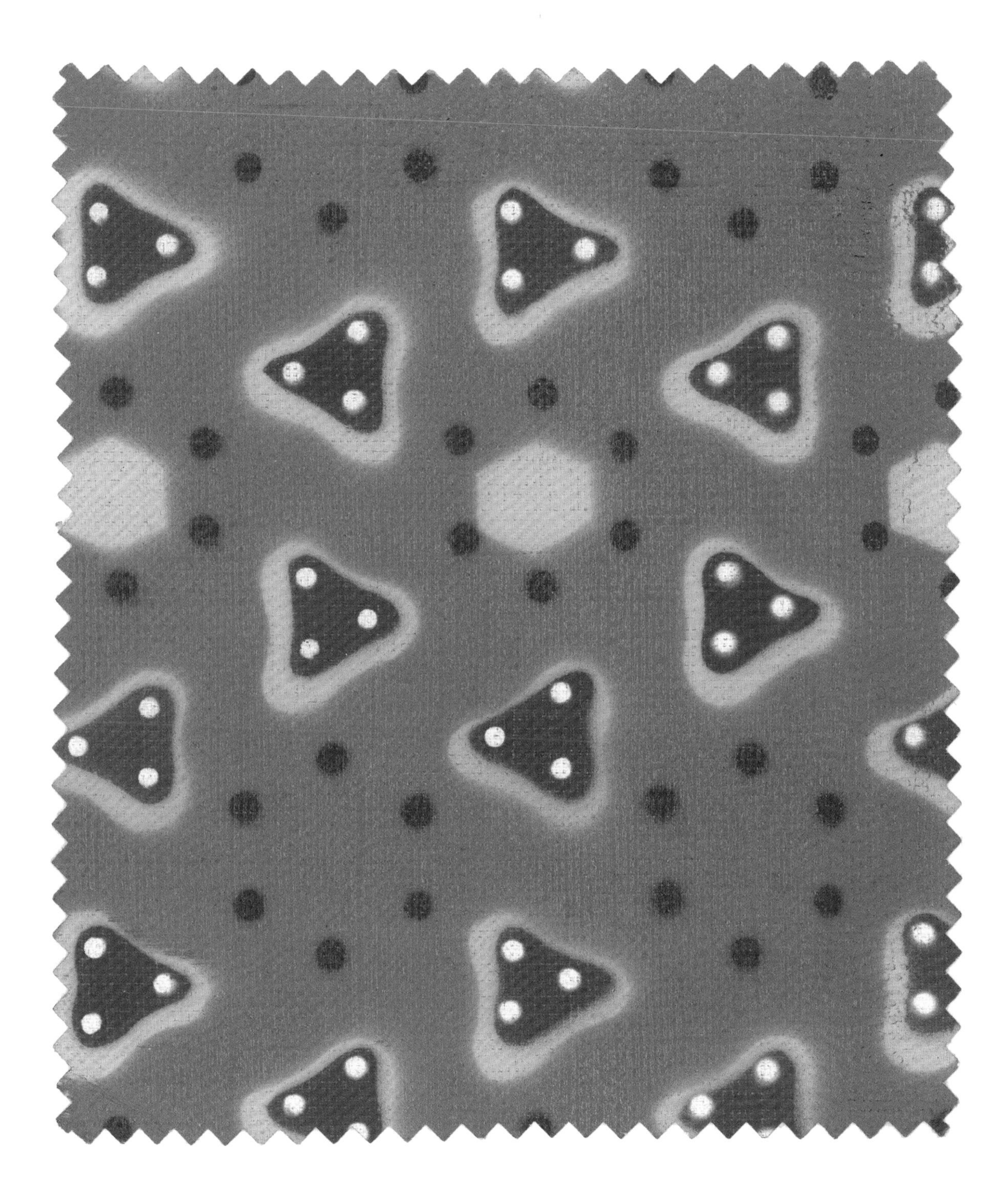

PRESERVED HUMAN SKIN WITH TATTOOS

Men in the military services have a strong tradition of injecting ink into their skin to create tattooed illustrations.

This is one of 300 tattoo specimens collected on behalf of the pharmacist, philanthropist and entrepreneur Henry Wellcome. They were acquired in 1929 from a Parisian surgeon known as Dr Villette, who worked in military hospitals and is said to have harvested these portions of inked skin from the bodies of dead soldiers.

Today the collection, storage, display and disposal of this sort of bodily material is carefully regulated in the UK under the 2004 Human Tissue Act. Key to this legislation is the issue of consent, which is a requirement for all bone and soft-tissue specimens, including skin, dating from the past 100 years. Legislation of this kind did not exist at the time these tattoos were collected from various bodies post-mortem.

The idea of medical specimens, such as organs preserved in formaldehyde or bones prepared as articulated skeletons, is that they work as teaching tools, communicating information about anatomy, physiology and disease. These specimens are intended as example body parts and usually come from an anonymous human body. Specimens of tattoos are rather different; through their style and subject matter, they start to speak of the identity of the person who once wore the skin they came from. Tattoos might have decorative, symbolic, commemorative, identification or ritual functions and because of this can offer insights into the lives of the individuals they once covered. They have been chosen to remember loved ones, mark milestones in life, or serve as souvenirs to significant experiences, and have been linked to belief systems, magic, criminal deviance, and medicine throughout history.

When it comes to the role of tattoos in the history of medicine, they have been associated with both curing and causing disease. For instance, ancient mummies have been recovered with tattoos placed on their bodies in ways that suggest they served a healing function. During the era from which this skin fragment comes, the act of tattooing itself was linked to the transmission of syphilis and other infectious diseases.

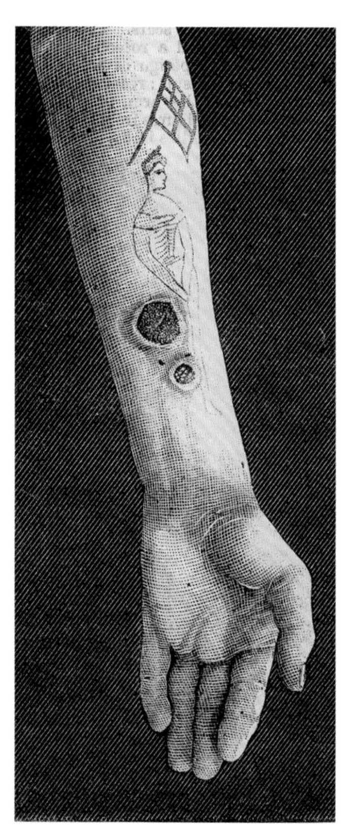

Left: An image showing syphilis at the site of a tattoo, 1889.

Opposite: Human skin tattooed with human faces and butterflies, French, 1880–1920.

While keeping human remains in museums, especially medical museums, is commonplace, it is also often problematic. Many museums hold remains that were collected during the colonial era under sinister circumstances. How to respond to this horrible history clearly demands care, respect and action. In the wake of the 2004 legislation, some museums have started to repatriate these ancestral remains, but there is still work to be done to navigate this difficult past.

MYOGLOBIN – THE FOREST OF RODS

Made from steel rods and Meccano clips, this model shows myoglobin's complex structure.

Gigantic in scale and amazingly complex, this model of the protein myoglobin reveals the structure of a molecule that plays a critical role storing oxygen in the body. The model became known as the "forest of rods" because of the dense number of shafts supporting the protein structure weaving among them.

The painstaking task of analysing the protein and building this complex model was carried out by Cambridge scientist John Kendrew. In 1953 he became the first person to decode the structure of a protein, building a crude outline of myoglobin in plasticine. Kendrew then began work on the "forest of rods", attempting to reveal the detail of the protein at the atomic scale. Kendrew used a technique called X-ray crystallography to create the model. Firing beams of X-rays at crystals of myoglobin, he recorded the patterns of the X-rays as they were reflected by atoms in the molecule. Analysing over 25,000 of these patterns using EDSAC, an early electronic computer constructed at Cambridge University, Kendrew began to deduce the protein's shape.

Understanding a protein's shape is fundamental to revealing the role it plays in the body. Model building was the only way for scientists at the time to map and understand the 3D structures of molecules, which are not easily conveyed in words or pictures. Inspired by Meccano, the children's construction kit, Kendrew began to fix Meccano clips of different colours to metal rods to plot the likely position of atoms according to his X-ray images. Layer by layer he revealed the contours of the protein's shape and the mechanism explaining how myoglobin was able to store a single oxygen molecule. Kendrew was surprised by how irregular the protein's shape was.

Kendrew won the 1962 Nobel Prize for Chemistry for determining the first atomic structures of proteins using X-ray crystallography, along with his colleague Max Perutz, who built upon Kendrew's work to identify the structure of a more complex protein, haemoglobin. Decoding the identity of biological molecules paved the way for better understanding of health and disease, and the development of new forms of diagnosis and treatments.

Left: John Kendrew adjusting his molecular model, 1957.

Opposite: Model of myoglobin known as "forest of rods", 1957–60.

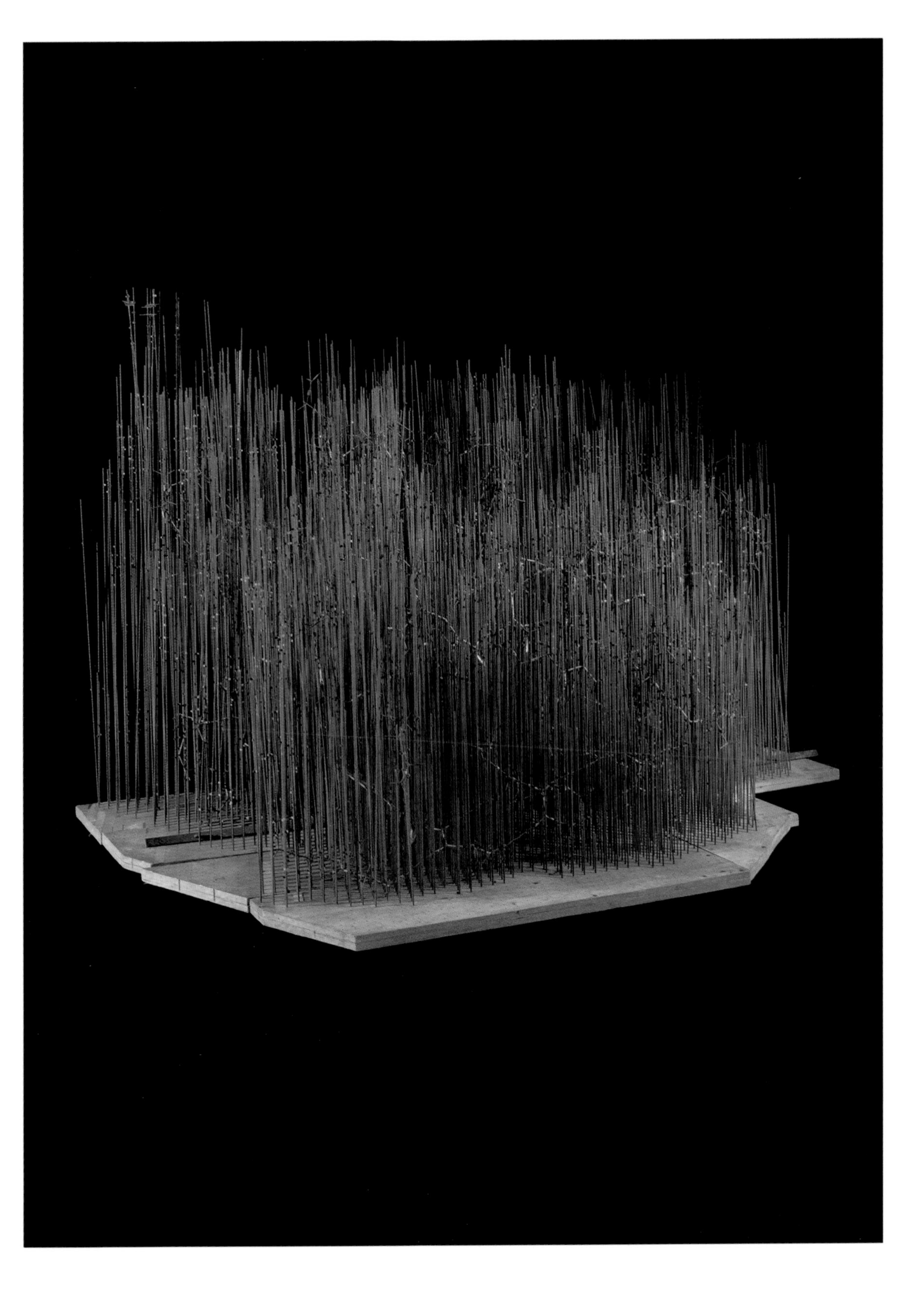

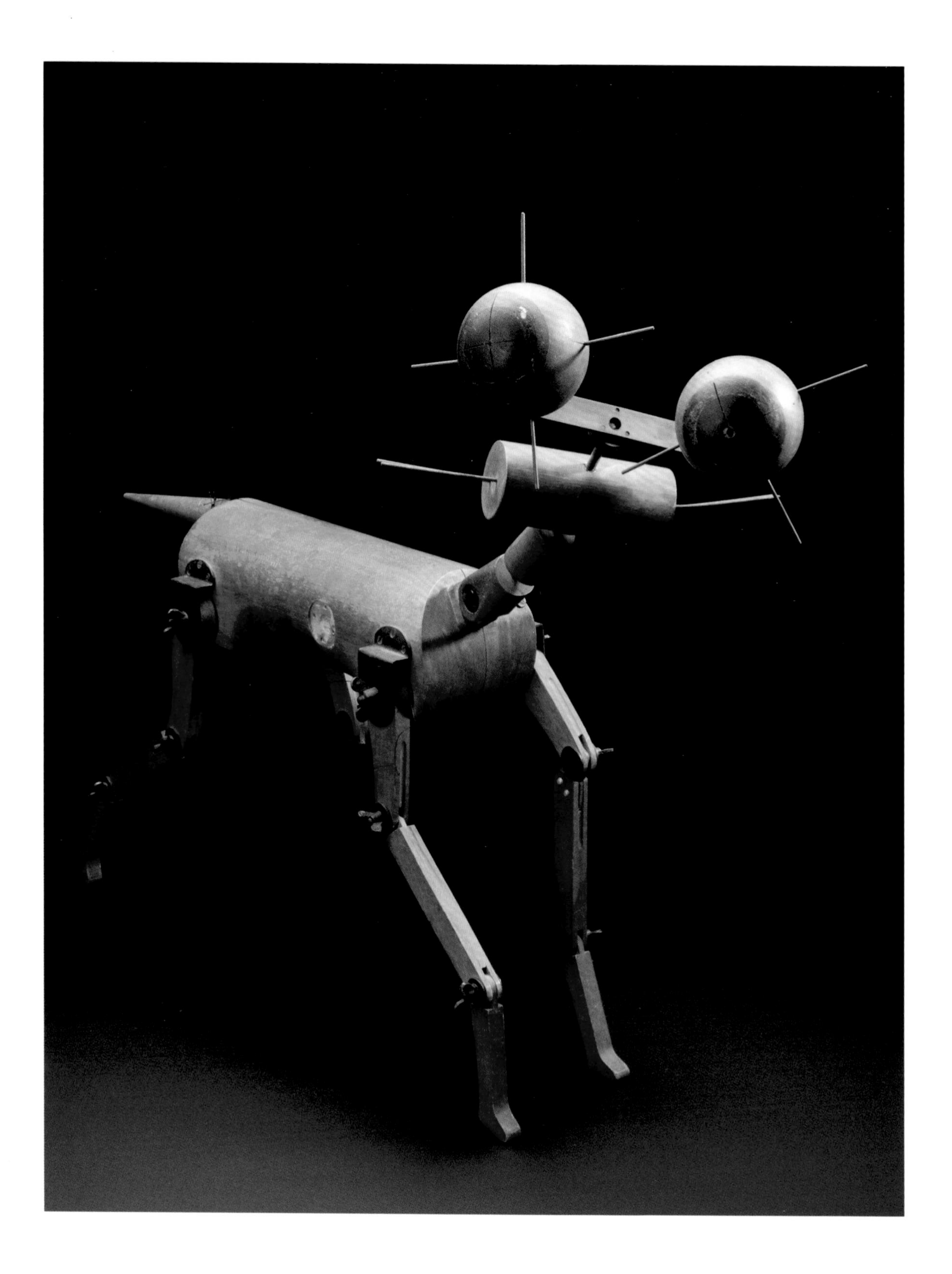

MODEL SHOWING A CAT'S REFLEXES

Scientists could examine wooden models of animals as an alternative to running tests on live animals.

This curious model of a cat was designed to demonstrate the action of the reflexes. It is thought to have been made for the British physiologist Sir Charles Scott Sherrington, who was renowned for his work in this area. Sherrington advanced scientific understanding about the reflexes and the function of neurons. He developed Sherrington's Law of Reciprocal Innervation, which describes the inhibition of muscles when their opposites are stimulated, and he was the first to use the word "synapse" to name the point where signals pass between nerve cells. So important was Sherrington's contribution to knowledge that he and his colleague, Edgar Adrian, were awarded the Nobel Prize in Physiology or Medicine in 1932.

Sherrington's work paved the way for modern understandings of neurophysiology. Yet it was reliant on the animals used in his experiments. Apes, dogs and cats all played their part and were often subjected to surgery to render them more suitable research subjects. In *The Integrative Action of the Nervous System* (1906), Sherrington describes how, to facilitate his research, cats and dogs had their spinal cords cut and significant portions of their brains removed while still alive, albeit under anaesthetic.

Vivisection, or subjecting live animals to surgical operations for the advancement of knowledge, has been used in medical research throughout history. During the eighteenth century figures such as Samuel Johnson and Jeremy Bentham voiced their concerns about animal suffering. Yet since the nineteenth century targeted opposition to the controversial practice of vivisection has mounted. The first anti-vivisection organization, the National Anti-Vivisection Society, was established in 1875 by the social reformer, activist and campaigner for women's suffrage Frances Power Cobbe. This was soon followed by the introduction of the Cruelty to Animals Act 1876, which stipulated that a licence was required to conduct animal experiments and that any testing on animals had to be deemed essential in pursuit of original research. In 1898 the British Union for the Abolition of Vivisection was founded. Sherrington's experiments took place in the wake of these reforms, when concerns about animal cruelty were gaining increased momentum.

Some want to see animal experimentation end due to the suffering it causes. Others argue that advancements in medical knowledge warrant experimentation on animals, but this is contentious – these practices are based on the understanding that human and non-human animal bodies respond and behave in the same way, which is not always the case. Perhaps this cat was produced as a more ethical pedagogical model in the face of these concerns?

Opposite: Model of a cat to demonstrate its reflexes, 1940s.

Above: An engraving by D.J. Tompkins, 1883, after a painting by J. McClure Hamilton, showing a dog apparently begging to be spared from vivisection in a laboratory. The image demonstrates the mobilization of anti-vivisection stances at the end of the nineteenth century.

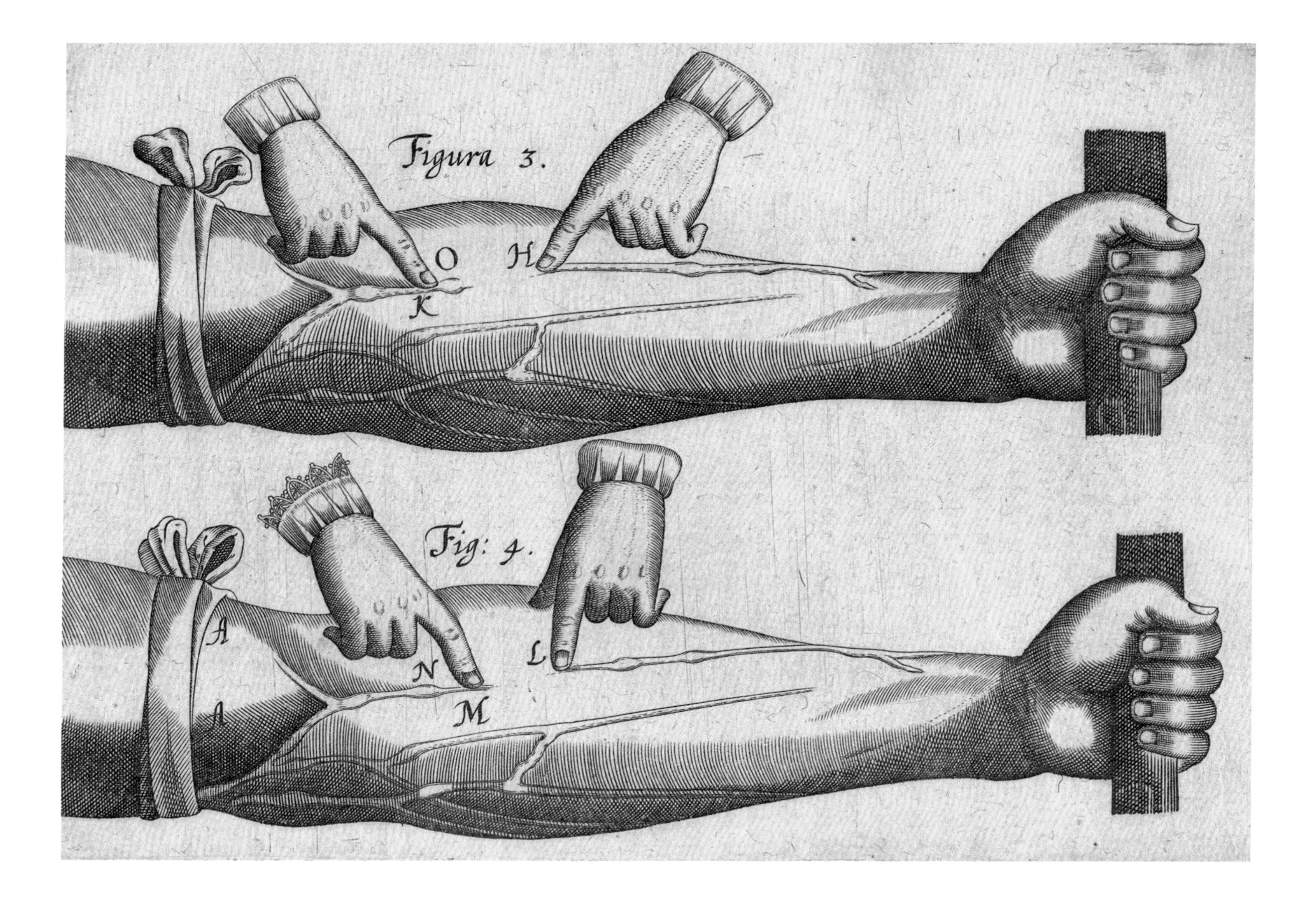
Figura 3.
O
H
K
Fig: 4.
A
L
N
M
A

HARVEY'S *ESSAY CONCERNING THE MOVEMENT OF THE HEART AND BLOOD IN ANIMALS*

William Harvey was the first person to describe the circulation of the blood around the human body, pumped by the heart.

It is obvious to us that the familiar thumpity-thump of our pulse reflects the heart circulating blood around our bodies. However, for thousands of years people had a very different understanding. They believed that the liver was responsible for producing blood for the veins, with the heart making blood for the arteries.

William Harvey, an English doctor, was the first person to challenge these ideas and accurately describe the circulation of blood by the heart in his book *An Anatomical Essay Concerning the Movement of the Heart and Blood in Animals*, published in 1628. Born in Folkestone, Harvey was educated at Cambridge University before studying medicine at the University of Padua in Italy, one of Europe's leading centres for medicine. Returning in 1602, Harvey established himself as a physician. He married well and began moving in elite social circles. In 1618 he was appointed physician to James I, then to his son Charles I when he became king.

Harvey's studies at Padua, with their emphasis on hands-on experimentation, dissection and observation, proved formative. In addition to dissecting human corpses as an anatomy lecturer, he began to conduct countless experiments on animals to understand the inner workings of the body. He observed the beating hearts of dogs and birds and measured the total amount of blood that could be drained from animals, comparing it to the volume their heart could empty with each beat. He calculated that new blood could not be generated in sufficient quantities according to traditional theories, but rather that the blood supply must be a fixed volume moved in a circuit instead. Other experiments were simpler but equally effective. Applying a tourniquet to the arm, he demonstrated that the presence of valves permitted blood to flow only in one direction, an experiment illustrated within his book.

Opposite: A page from *An Anatomical Essay Concerning the Movement of the Heart and Blood in Animals* showing one of Harvey's experiments, published 1628.

Above: William Harvey demonstrating his theory of circulation of the blood before Charles I, by Ernest Board, 1920s.

Though Harvey had shared many of his observations in public lectures, in 1628 he finally published his ideas after 10 years of painstaking research. His book received a mixed reception. Many people found it hard to accept his ideas as they contradicted important medical theories of the time, including the rationale for bloodletting. Harvey's theory of the heart as a pump took more than half a century to become accepted medical knowledge, and its wide adoption was something he did not live to see.

LEEUWENHOEK'S MICROSCOPE DESIGN

Antonie van Leeuwenhoek used his early version of a microscope to discover and view an invisible world.

The 1600s saw the unfolding exploration of a new world – not a new land but of a world ordinarily hidden to the naked eye. This small brass device is a replica of the microscope invented by Dutch businessman and scientist Antonie van Leeuwenhoek, one of the first people to explore the strange and magical world at microscopic level and to observe certain cells and bacteria for the first time. Raised in Delft in the Netherlands, van Leeuwenhoek worked as a textile merchant. Accustomed to using magnifying glasses to examine the quality of the textiles he traded in, van Leeuwenhoek developed an interest in glass lens making. He developed a new technique for making lenses that could magnify specimens up to 200 times – a giant leap forward compared to other microscopes at the time, which were only able to magnify objects to about 20 or 30 times their natural size.

His miniature microscope consisted of one of these lenses fixed between two brass plates, behind which sat a small screw. A specimen could be placed on the tip of the screw, which could be turned to adjust how close the object was to the lens and bring it into focus. Moving the microscope close to his eye and looking through the lens under a bright light, van Leeuwenhoek could examine the specimen. Driven by insatiable curiosity, van Leeuwenhoek examined almost anything that could be placed in front of a lens: blood, sperm, pond water and even the plaque between his teeth. Van Leeuwenhoek frequently reported sightings of "many very little living animalcules, very prettily a-moving" – the first observations of micro-organisms. Not being able to draw well, he hired an illustrator to prepare drawings of the things he observed to accompany his careful descriptions.

Van Leeuwenhoek did not write any books, but from 1673 he began to share his research with the Royal Society in London. Though they were sceptical at first, van Leeuwenhoek's work gradually captured the attention of the Royal Society's members, who eventually published his microscopic observations on mould, bees and lice in their journal. On his death van Leeuwenhoek left 247 microscopes and 172 lenses. However, only nine of his microscopes have survived. Replicas like this one have helped modern researchers understand and learn from van Leeuwenhoek's techniques.

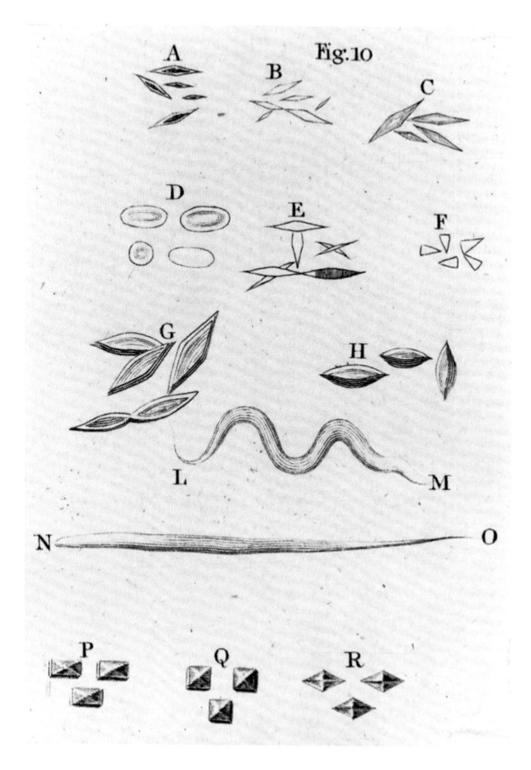

Above: The microscope animalcules seen by Antonie van Leeuwenhoek in white wine vinegar, 1799.

Opposite: Model of van Leeuwenhoek's microscope, 1901–30.

HERPES ZOSTER

Frontal & trochlear branches of fifth nerve

ILLUSTRATION OF WILLIAM WOOLGAR

Artist Edwin Burgess created striking and technically accurate portraits of diseases of the skin.

William Woolgar was three years old when he experienced a severe episode of herpes. Generally a healthy child, he started complaining of headaches on a Wednesday and by Thursday a severe rash had appeared, spreading from his right eyelid all the way to the top of his forehead. After several sleepless nights he was taken to see a doctor, which was when this portrait was made by artist Edwin Burgess.

Burgess had been commissioned by Jonathan Hutchinson, Secretary of the New Sydenham Society, to produce an *Atlas of Skin Diseases*. Hutchinson was an exceptionally gifted clinical diagnostician, renowned for the accuracy and logic of his observations made by the naked eye. He recognized the crucial role of detailed and accurate medical illustrations in training physicians to spot, recognize and distinguish one condition from another. Unlike other medical illustrations of the time, which allowed some creative licence to the artists, Hutchinson insisted that these new portraits be solely based on observation and free of any artistic interpretation and rearrangement. As he put it, "the *Atlas* was, therefore, not only illustrations of typical varieties of diseases, but faithful portraits of individual patients." Indeed, Burgess devoted as much care to the representation of William's lace collar as he did to the vesicles on his forehead, although only the latter were relevant to clinical diagnosis.

Reinforcing the idea of these illustrations being portraits of individuals and not only of diseases, Hutchinson compiled an accompanying book to the *Atlas* in which he gave the patient's story alongside the medical description of the ailment. It is here that we discover William's name, age and personal experience of dealing with herpes. We do not know the circumstances of how he came to contract the disease but we feel for him as we learn that he cried in his sleep from the pain. These portraits and accompanying information are truly human, such was the strength and novelty of the New Sydenham Society's *Atlases*.

Although these illustrations were intended as diagnosis and teaching tools, and the Society prided itself on pure scientific objectivity, they also capture the vulnerable moment when the patient is scrutinized by the physician – and in this case by the artist, too.

Opposite: Medical illustration showing herpes zoster by Edwin Burgess, commissioned by the New Sydenham Society, 1860–84.

Right: Portrait of Jonathan Hutchinson, clinician and Secretary of the New Sydenham Society, by Leslie Matthew Ward, 1890.

A MEASURE FOR HEIGHT

James Tanner used specially designed tools for measuring children in the Harpenden Growth Study.

Were you tall or short as a child? This giant ruler, known as a stadiometer, was used to measure the height of children in a ground-breaking study of child growth and development carried out between 1948 and 1971 by Professor James Tanner and his colleague Reginald Whitehouse.

In 1948 Tanner was invited by the British Government to take over a study of childhood growth, begun during the Second World War, that had been observing the effects of wartime malnutrition on growth. Tanner and Whitehouse gathered detailed measurements of 450 boys and 260 girls living in a children's care home in Harpenden, southern England. Every child in the study was measured regularly from the age of three or four. Fifteen different measurements were taken on more than 9,000 occasions, to create a record of a child's physical changes from childhood to early adulthood. The Harpenden study also gave rise to what came to be known as the Tanner scale, a checklist of features that define different stages of puberty. Theirs was one of the first studies to follow and measure how a generation of children grew and developed between pre-school and early adulthood.

Above: Professor James Tanner, *c.*2000.

Opposite: Stadiometer, 1965 (used by James Tanner, Harpenden Growth Study).

The stadiometer recorded the height of a child with a sliding wooden panel, which rested on top of the child's head and indicated a point on a scale running down the ruler. Tanner made detailed charts tracking the growth of these children over many years, which he used to define normal patterns of development. The charts he created enabled a child's growth rate to be plotted against the average over time; data used by doctors and scientists around the world to monitor whether a child's development was healthy. Tanner showed that there is not one normal pattern of growth in adolescence but several: early, middle and late developers. His research on height also showed relationships between growth and the children's physical and social environment. He remarked: "A child's growth rate reflects, better than any other single index, his state of health and nutrition, and often indeed his psychological situation."

Tanner was also among the first scientists to study the use of human growth hormone. In 1956 he set up a pioneering research centre for child growth at the Institute of Child Health in London, working with groups of young patients to treat them for delayed growth. The hormone, which could only be extracted at that time from human cadavers, tragically led to several deaths through the transmission of an infectious brain disease, forcing Tanner's trials to stop. Treatment resumed in the 1990s, when genetically engineered human growth hormone was introduced.

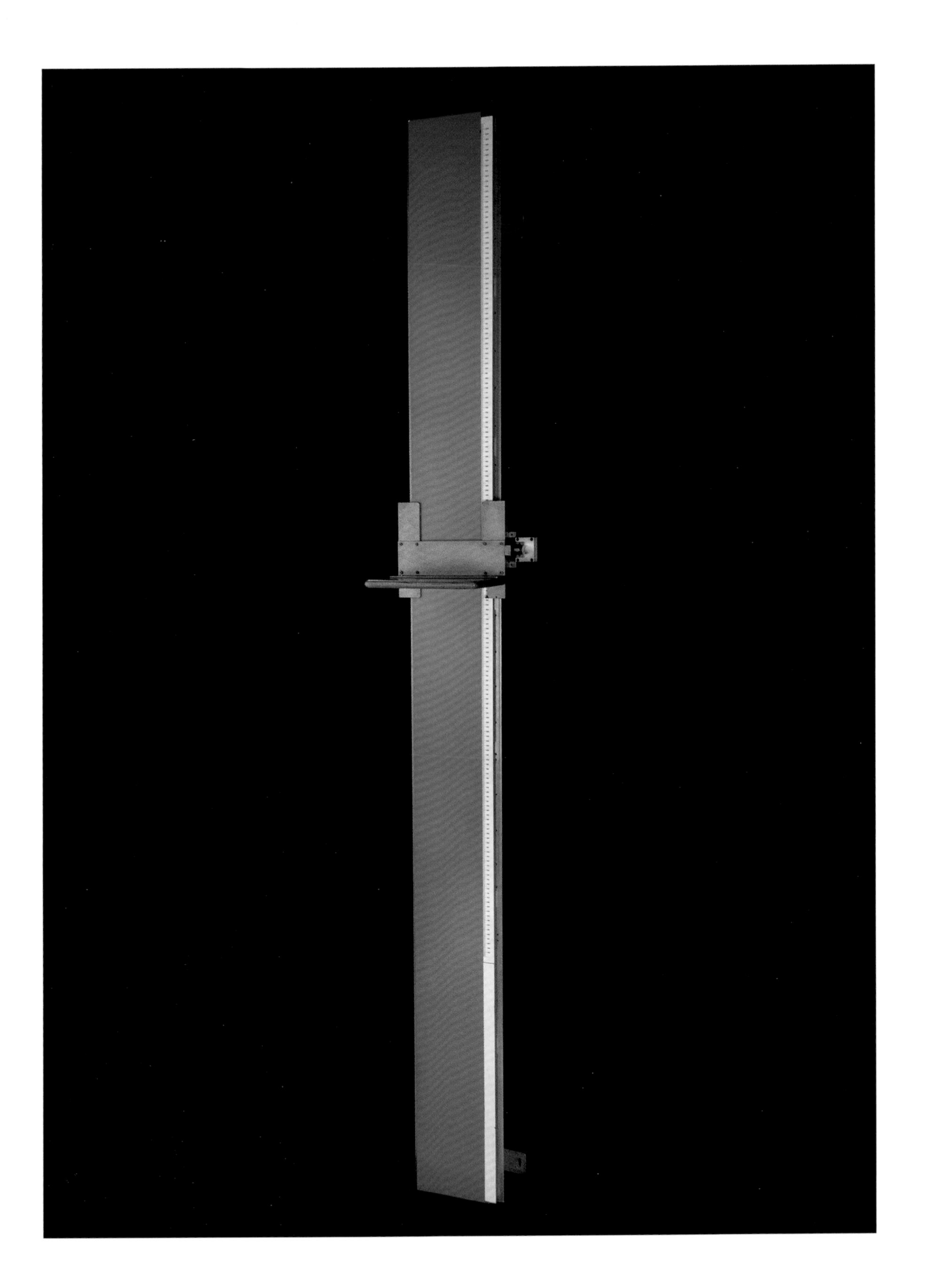

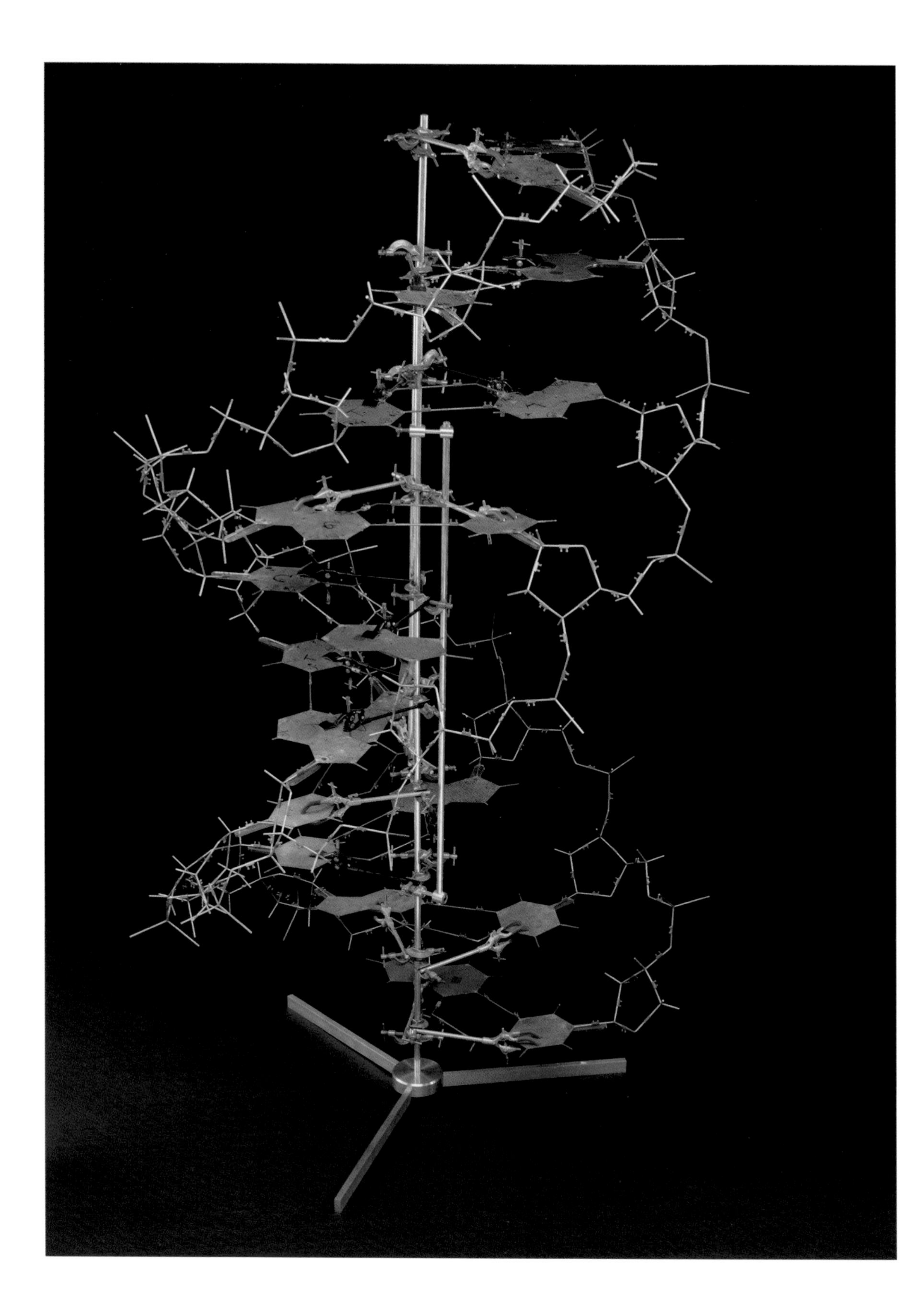

RECONSTRUCTION OF CRICK AND WATSON'S DOUBLE HELIX MODEL OF DNA

The solution to the structure of DNA is demonstrated in this model, clearly showing the now-famous double helix.

Scientists need to promote even the best ideas, and successful communication is part of their achievement. This model is a legacy of the process by which two young men turned an idea into a legend. It has proved to be the most memorable thing associated with the birth of molecular biology, a science that links molecules to life.

In 1953 Francis Crick and James D. Watson, of the Cavendish Laboratory in Cambridge, suggested a structure for the molecule known as DNA. This chemical was already being credited as a crucial agent in the process by which life reproduced. Crick's and Watson's proposal of a double-stranded helix structure fitted evidence of the structure from the X-ray crystallography of Rosalind Franklin and Maurice Wilkins at King's College London. It also suggested how the molecule, and crucially living creatures, could store information and replicate.

A scientific article was published in the journal *Nature* in April 1953. But these young men still needed to convince their peers and, shortly after their insight, they had a large model constructed. This was photographed for an article in *Time* Magazine – a photograph which was not then used but which would become iconic in later years. The model itself was displayed at an open day at the Cavendish Laboratory, then put aside. In the late 1960s, when the scientific import of the discovery was becoming clear, Watson wrote a best-selling account of the discovery of the DNA structure and included a photograph of this model. This book made the discovery of the double-stranded helix structure of DNA the defining event in the history of the new science of molecular biology.

The model itself was neglected and dismantled. In the 1970s staff at the Science Museum identified some of the original bases and commissioned Dr Farooq Hussain to reconstruct the original, supplementing original parts with newly made replacements for missing components. Some parts, therefore, have been made by the Science Museum, but other components of this model date back to the early part of 1953, when Crick and Watson were promoting their radical idea.

Opposite: Reconstruction of the double helix model of DNA using some of the original metal plates, by Francis Crick and James Watson, England, 1953.

Above: James Watson and Francis Crick with DNA model, signed by Watson and Crick, by photographer Anthony Barrington Brown, original 1953, reprinted 1993.

JAPANESE ACUPUNCTURE MODEL

Wooden models of the human body provided a map of the meridians of energy used in acupuncture.

Long multi-coloured lines dotted with Chinese characters stretch across this wooden figure, connecting different parts of the body. Although almost impossible to decode to the untrained eye, these lines helped practitioners to visualize the invisible channels or meridians through which *qi* (or *chi*) is said to flow and the various points where acupuncture needles can be inserted.

In Traditional Chinese Medicine and in Japanese medicine, known as Kampo, acupuncture is intended to encourage the flow and balance of *qi* throughout the body. *Qi* is understood as a vital life force that transforms and affects the human body in a variety of different ways. Disharmony between different types of *qi* is believed to cause ill health, and so acupuncture, along with a variety of other interventions, is used to restore harmony. By inserting needles in particular points along the channels, acupuncture specialists believe they can assist the flow of *qi* and encourage healing.

This Japanese model quite literally embodies this understanding of the human body. Chinese characters (which are also used within the Japanese language) have been inscribed by hand across the entire body, showing us where the needles should be inserted. In some places these characters are written upside down, showing us the direction in which *qi* is thought to flow. Looking at the figure in more detail, we can also see that the channels are all painted in different colours: green, red, yellow, white and blue/black. These colours represent the five phases of transformation that *qi* is believed to undergo in the human body: wood, fire, earth, metal, and water. Within the Five Phases Theory, these phases or elements interact in different ways, having significant impact on different organs, tissues, senses and emotions. To influence these interactions and encourage harmony, acupuncture needles would need to be inserted in specific points along these channels, depending on the patient's symptoms.

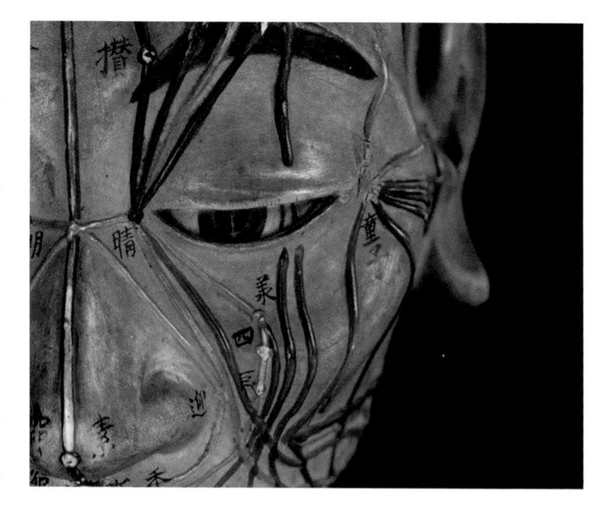

Above: Detail of head showing the meridians, or channels through which *qi* (*chi*) or energy flows in the body.

Opposite: Wooden acupuncture model, Japan, 1681.

From an inscription in Japanese on one of the model's shoulders, we know that it was either made or owned in 1681 by someone with the surname Ishihara. It is one of many models that were created during the Edo period in Japan (1603–1868). At this time acupuncture specialists were eager to improve the training that students received and ensure that they practised acupuncture with precision. These models, studied alongside textbooks or manuals that decoded this complex web, made visualizing the channels far easier, and helped students to put the theories that they had learned into practice.

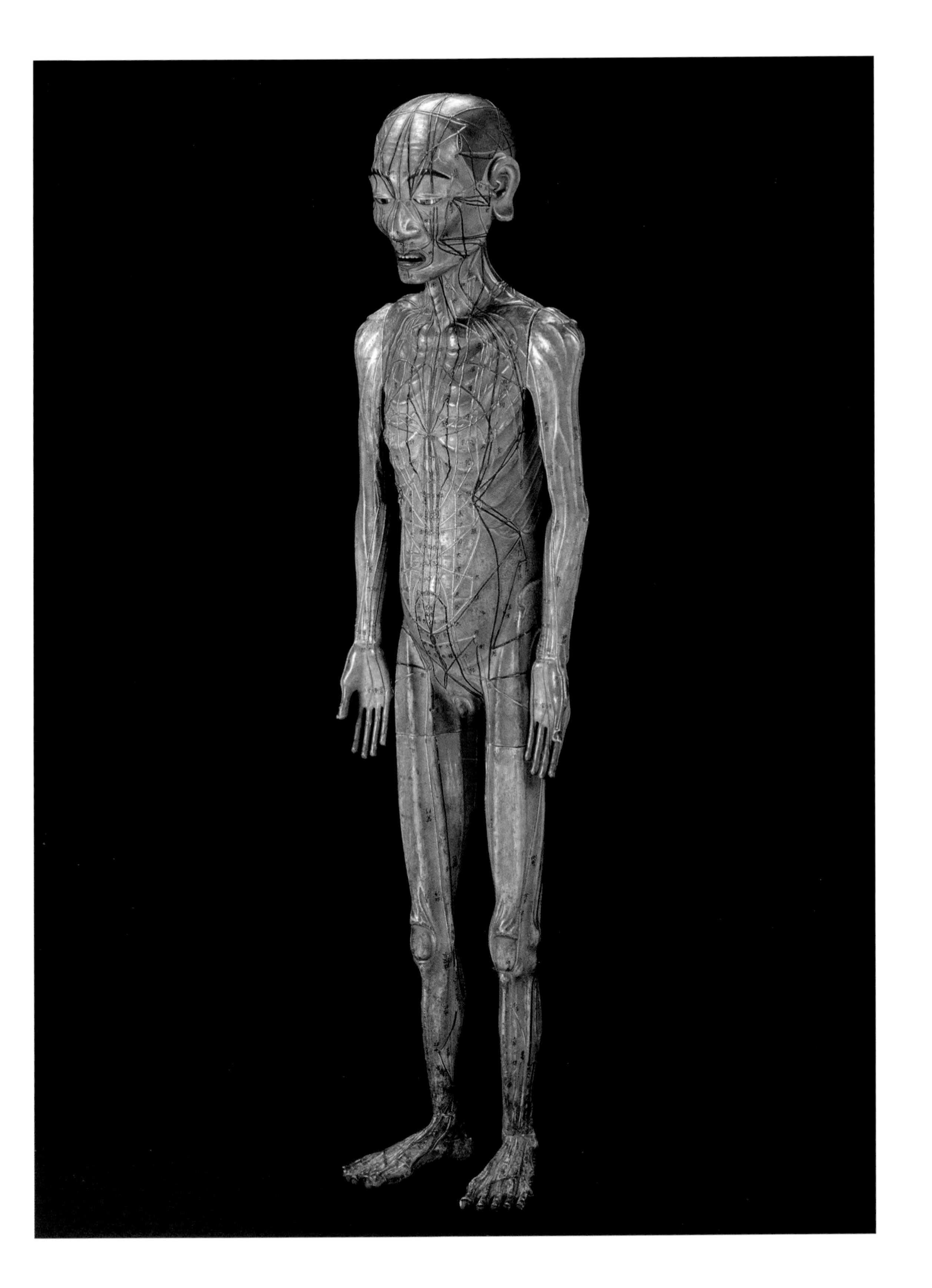

ABI
PRISM
3700 DNA Analyzer

DNA ANALYZER

DNA analyzers played an essential role in the Human Genome Project, which aimed to identify all human genes.

Deoxyribonucleic acid, more commonly known as DNA, is the instruction manual to build organisms and keep them alive. This complete set of instructions is also referred to as the genome. DNA analysis has become mainstream and ubiquitous – it is used by doctors to inform medical diagnoses, by pharmaceutical companies to develop more targeted and efficient drugs, by investigators to identify suspects and victims, and by members of the public to research their ancestry.

None of this would be possible without the Human Genome Project, an international scientific collaboration whose aim was to identify all human genes. For years DNA analyzers across the world worked non-stop towards sequencing the whole human genome, which involved determining the order of the four chemical building blocks that make up DNA – namely adenine, cytosine, guanine and thymine. The ABI PRISM® 3700 DNA Analyzer, pictured here, played a pivotal role in decoding the three billion letters that make up human DNA. Developed in 1998 by PE Biosystems, it was designed to process an unprecedented amount of data, and produced results superior to any other existing machine, at a lower cost. The increased throughput, automation and cost-effectiveness made this device the tool of choice for DNA analysis. The Human Genome Project was successfully completed in 2003, more than two years ahead of schedule; an impressive performance partly thanks to the incredible technology of these new devices.

The new information revealed by the project helped scientists understand how variation in the human genome can impact health and cause diseases. The very first sequencing of the human genome took 13 years and cost $2.7 billion, nearly $5 billion (£3.8 billion) today. Fifteen years later, thanks to continually evolving technology, genomes can be sequenced in a matter of weeks for a few thousand dollars. This means that more information is constantly discovered and released, shaping the future of medicine and diagnosis. Understanding how certain gene mutations can cause diseases helps doctors form a quicker diagnosis and directs scientists in their research for more effective and targeted treatments.

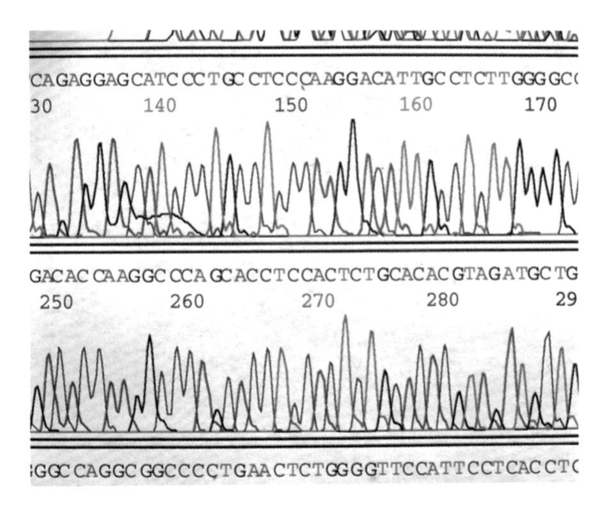

Opposite: ABI PRISM® 3700 DNA Analyzer by PE Biosystems, 1999.

Above: A chromatograph from the human genome mapping project carried out by the Medical Research Council in Cambridge.

WAX ANATOMICAL MODEL

Fragile, anatomical models made of wax present a historic challenge for those who work to preserve them.

The Science Museum holds a series of seven wax figures, which likely originate from the workshops of Clemente Susini at La Specola in Florence. Waxes were produced here in the eighteenth century for medical training, artist instruction and education of the public. This series of anatomical waxes represented the functional relationships in a healthy, living body. Waxes were generally made using plaster moulds into which several layers of molten wax were poured. The wax was either coloured or painted on the inside, with finer details imitated using textile or metal threads covered in wax and paint. When complete, the waxes were covered in a shining varnish, both to resemble the moistness of living tissues and to protect the wax from dust. Larger waxes were normally constructed around an armature for structural stability.

Wax as a material has advantages and disadvantages from a conservation perspective. It is very stable, it does not discolour and if its colour does change, this is usually due to pigments fading or a surface coating deteriorating. The disadvantages are that it is soft, can attract dirt and dust that stick firmly to the surface, is easily broken, and can be subject to shrinkage, which often results in hairline cracks.

When cleaning the wax sculptures, modern-day conservators must take care not to use any solvents that will change the varnish, wax, paint or pigments. Analysis of the varnish in another figure in the same series confirmed it to be pine resin, probably applied at the time of production. Although resin layers have been removed from similar models in the past, because of their tackiness, it was decided to leave the resin layer intact to preserve as much of the original look as possible. The figure was cleaned with small amounts of de-ionized water and soft brushes. One of the fingers was re-attached using a conservation adhesive and an entomology pin as support. This wax figure was X-rayed to assess the structure and condition of the internal armature and the thickness of the wax, and to view cracks and old repairs. Some of the dowels seen in this X-ray are repairs, consistent with points where cracks have appeared. It is incredible that these intricate objects made from a material as soft and fragile as wax have survived relatively intact for so long and can still be used for inspiration and study.

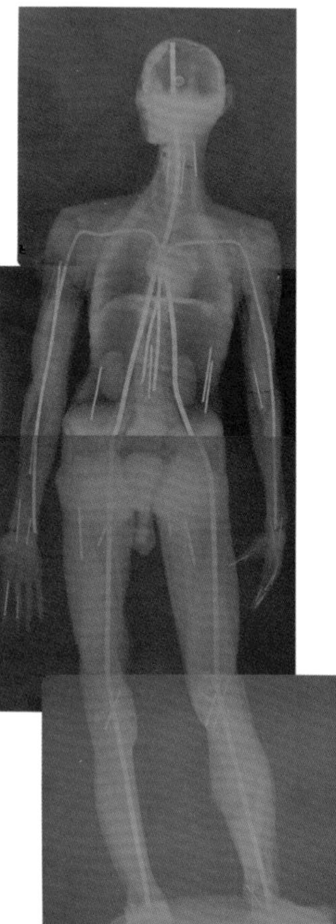

Right: X-ray of wax figure showing internal armature and repairs.

Opposite: Wax male anatomical figure showing muscles, arteries and veins in the body, 1776–80.

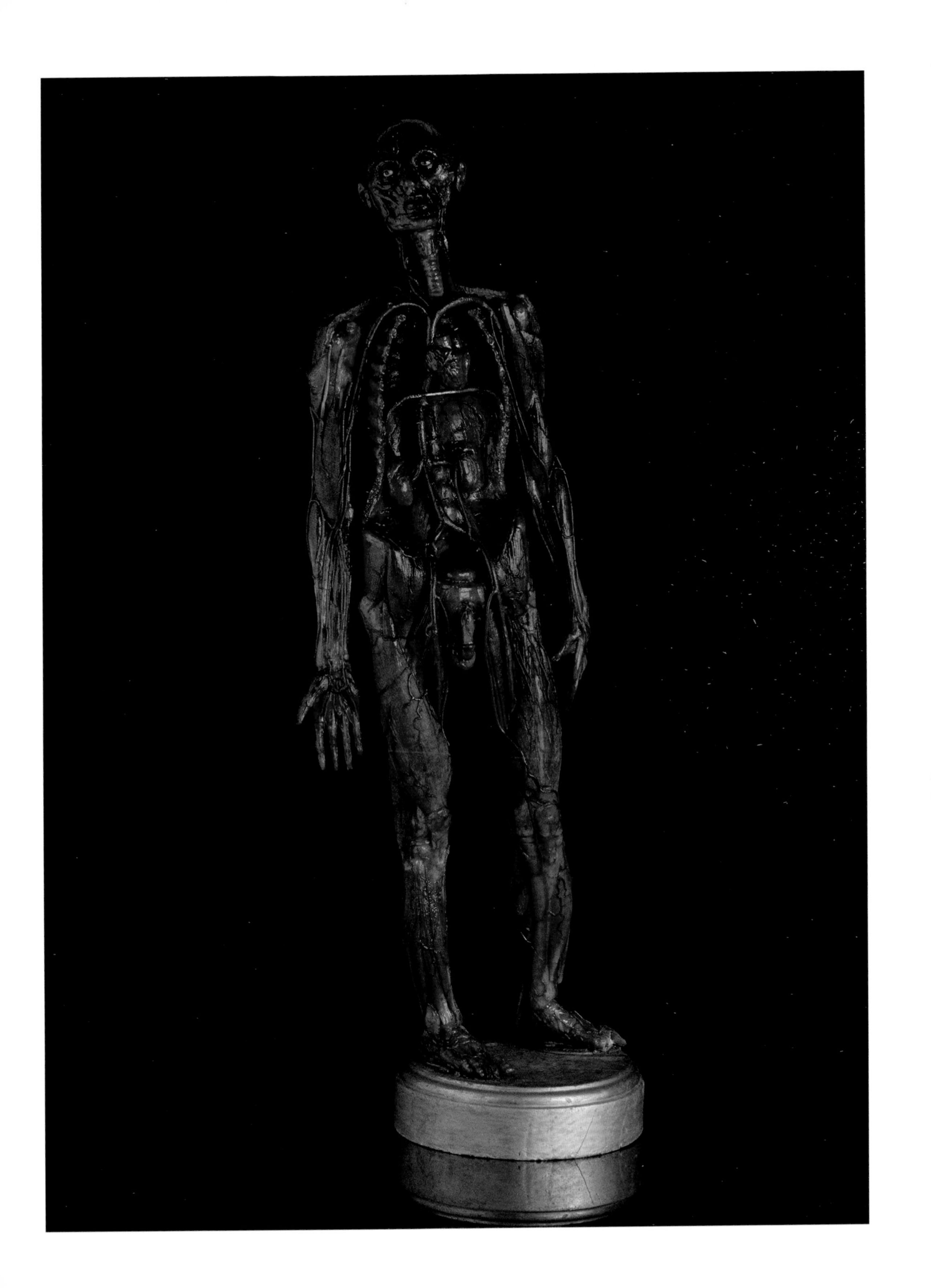

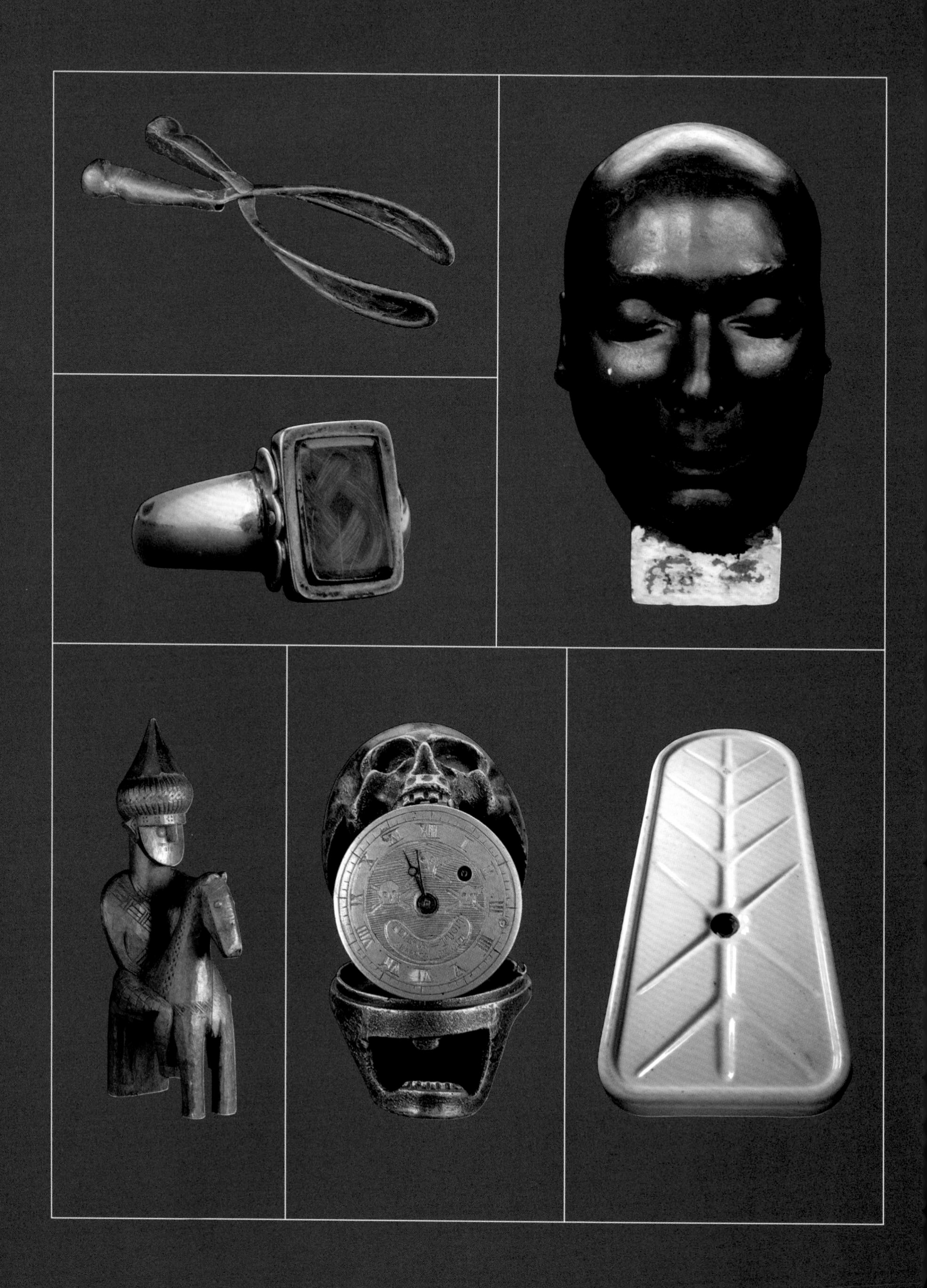

2

BIRTH AND DEATH

Examining the health of a human being can begin months before their birth and continue long after their death. A detailed examination of the interior and exterior of the body after death, known as a post-mortem, is one of the most effective ways to learn the cause of death. Although using dead bodies has helped improve medical knowledge of the human body and its conditions, such a practice raises questions about the ethical use of human remains. How the dead are treated and remembered is as complex as how we treat human beings in their earliest stages of life. Caring for a woman before, during and after childbirth is a tradition that often has a political ideology behind it, especially around areas such as pain relief and medically assisted labour. From our first moments to our last, the human body has a significance beyond the purely personal.

MIDWIFE'S BAG

A travelling midwife's most important piece of kit was her bag, which contained all that was needed to assist with childbirth.

Though women have provided care and support for other women during childbirth for thousands of years, legal recognition and regulation of midwifery was only established in the twentieth century. In the UK, the 1902 Midwives Act introduced compulsory training and supervision for the first time. Before this, it was common for midwives to have no formal education; instead, they gained their skills through working with more experienced practitioners. The Act also meant that midwives were only legally allowed to facilitate low-risk births and were expected to call a doctor for challenging or problematic cases. Previously, when not training others, midwives often worked alone, visiting expectant mothers in their homes, delivering babies and providing postnatal care.

Midwives used kits such as this during the 1940s, when assisting with home births. Hefty in weight, this example folds open to reveal an inner canvas bag containing medications, dressings, instruments and a measuring glass. It features an integral metal sterilizer for immersing instruments in boiling water to reduce the risk of infection.

Infection is just one of the many risks involved in childbirth, which remains a leading cause of death among young women in economically developing countries. Infection, the most common of which is puerperal, or childbed, fever, is often introduced by unclean hands or equipment. From the seventeenth to the nineteenth centuries, many cases of childbed fever were actually caused by doctors themselves, who were unaware of the causes of infection. Sterilizing equipment, alongside the introduction of antibiotic drugs, helped maternal death rates fall in the UK during the 1930s and 1940s.

With new technologies and medicines childbirth became increasingly medicalized during the twentieth century. Physicians began to note the benefits of a hospital birth, and a report by the Maternity and Midwifery Advisory Committee in 1967 recommended that all babies be delivered in hospitals. However, the mid-twentieth century saw women's groups in Britain campaign against the "over-medicalization" of childbirth, and debates around the changing practices of maternity care were vigorous. Today women can choose to give birth at home, in a centre run by midwives, or in a hospital, and though technological and medical advances in childbirth continue, midwives remain a vital part of antenatal care.

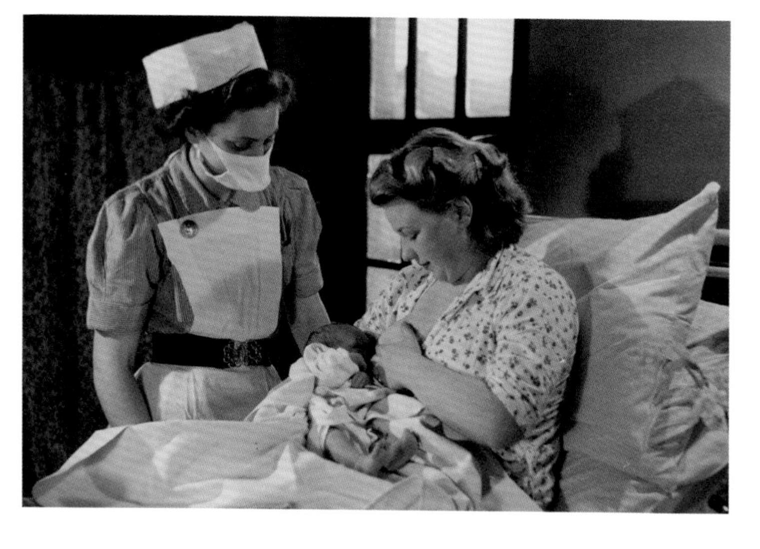

Left: Midwife providing instruction in breastfeeding *c.*1940.

Opposite: Combined box-type metal midwifery bag and sterilizer with some contents, made by W.H. Bailey and Son, London, 1940–48.

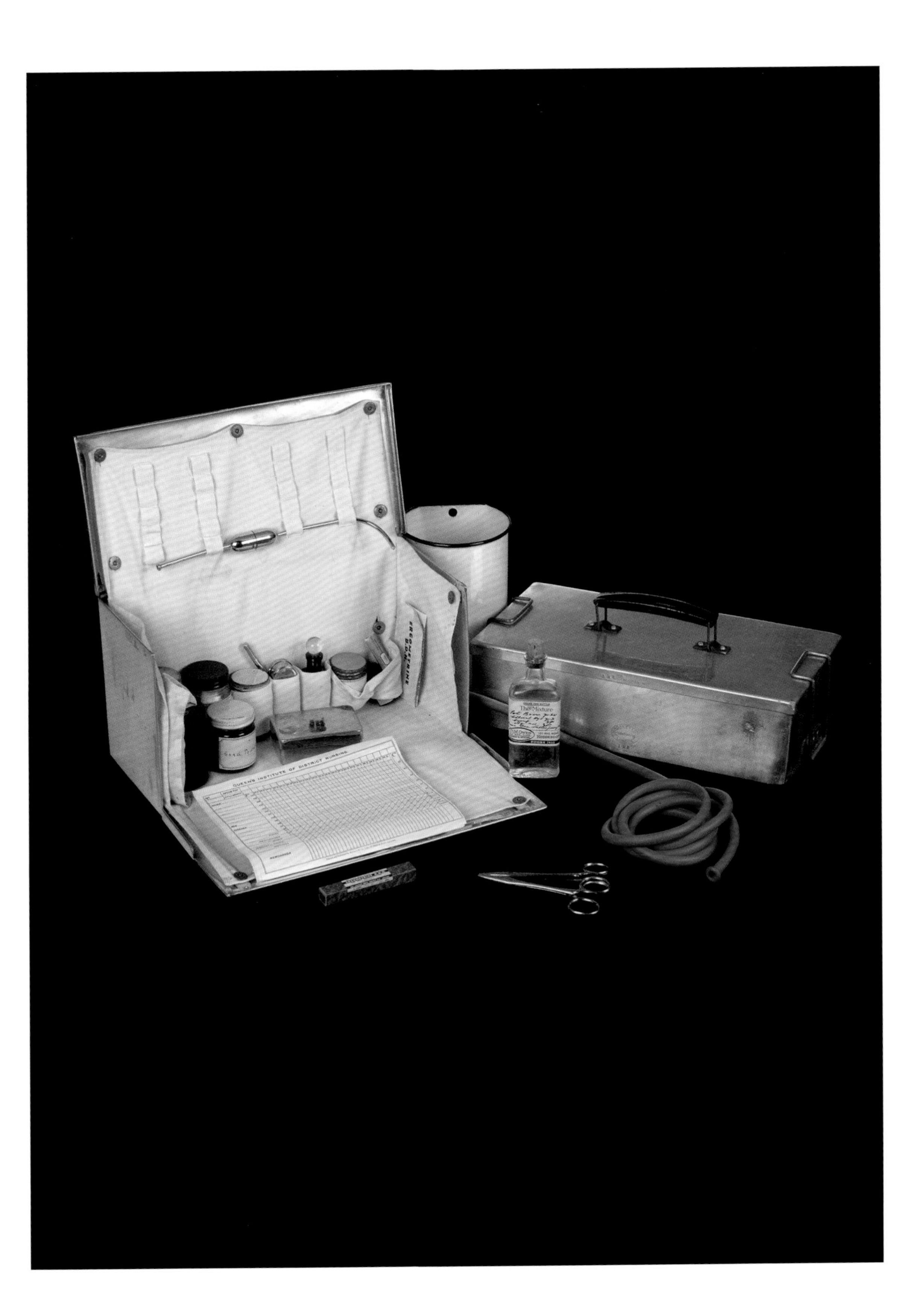
ERGOMETRINE
B.D.H.
QUEEN'S INSTITUTE OF DISTRICT NURSING
The Mixture

MEMENTO MORI REMEMBERING DR JOHN COAKLEY LETTSOM

In remembrance of the physician John Coakley Lettsom, this gold mourning ring contains a lock of his hair.

There are many ways to remember loved ones who have passed away. Memento mori jewellery, usually in the form of rings and lockets, was particularly fashionable between the seventeenth and nineteenth centuries. Latin for "remember you will die", memento mori were originally intended as reminders to the wearer to live a good, moral life in accordance with the scriptures.

Through the centuries the use and purpose of these pieces has evolved. This particular example is an item of mourning jewellery. The ring, which contains a plaited lock of hair from Dr John Coakley Lettsom set behind glass, bears an inscription to his memory and is dated 1815 on the back of the bezel. There is no information about the nature of the relationship between the wearer and the deceased, but Dr Lettsom was a well-loved and respected man, so many options are possible.

Born in 1744 in the British Virgin Islands, Lettsom was educated in England, where he trained as a physician. He was quickly regarded as the busiest, most philanthropic and most successful physician of his day, earning as much as £12,000 a year – some £2 million ($2.6 million) today – even though a large part of his practice was free of charge, and he gave away immense sums in charitable contributions. In 1773 he founded the Medical Society of London, a unique forum for experts from various medical branches to meet as equals and share knowledge and good practice. The Society, under Lettsom's impetus, enthusiastically supported and promoted the work of Edward Jenner, who was working on a vaccine for smallpox, one of the greatest killers of the time; he succeeded in 1796. After Lettsom's death in 1815, his friend and colleague Thomas Joseph Pettigrew was faced with the daunting task of writing his memoirs. Pettigrew opened by saying "to detail his useful and eventful life is to pronounce his highest eulogy".

Family, friends, patients and the medical community as a whole had reasons to commemorate Lettsom's life and death, but the presence of a lock of hair from the deceased suggests that this ring was made for someone personally close to him.

Opposite: Memento mori ring containing of lock of hair from Dr John Coakley Lettsom, 1815.

Left: John Coakley Lettsom, physician, with his family, in the garden of Grove Hill, Camberwell, *c.*1786.

TOBACCO RESUSCITATION KIT

Using tobacco smoke as an enema was an interesting way to resuscitate someone after drowning.

If you were unfortunate enough to find yourself drowning in the eighteenth and early nineteenth centuries, your rescuer might have used this equipment to revive you, quite literally blowing tobacco smoke into your rectum.

Concerned by the number of deaths resulting from drowning (and also the number of people mistakenly thought to be dead and prematurely buried), charitable organizations known as Humane Societies were established across Europe to promote the resuscitation of "persons apparently dead". Soon the rescue and resuscitation of drowned persons became a matter of public concern and responsibility. Founded in 1774, the London Humane Society encouraged research into methods of resuscitation and established new "receiving houses" where the victims of near-drowning could recuperate. They also deposited portable kits of resuscitation equipment along the river Thames, which could be used to speedily revive their patients.

Complete with a set of bellows and a fumigator, this nineteenth-century resuscitation kit is very different from the first-aid kits and defibrillators we use today. As you might expect, the bellows were used to inflate the lungs and help the patient to resume breathing (mouth-to-mouth was a familiar technique but fears over the spread of disease meant that the bellows were often preferred). However, they also had a second, very different function. Used with the fumigator, in which tobacco leaves would be burned, the bellows were pumped to deliver a tobacco smoke enema into the rectum.

Why tobacco? Warming the body was considered key to resuscitation, leading physicians to insert different "stimuli" into the body to encourage the heart to resume beating. According to the theory of the four humours, the four bodily fluids known as phlegm, blood, yellow bile and black bile needed to be in balance to maintain good health. Within this framework the warm and dry vapours created by burning tobacco and nicotine would revive recently drowned victims, who were believed to suffer from an excess of wet phlegm and cold black bile. It was believed that directing these fumes into the bowels would encourage a strong and regular heartbeat.

Although doubts about the effectiveness of this technique were expressed from the end of the eighteenth century, a small fumigator was nonetheless included in this kit. Clearly the London Humane Society preferred to be safe rather than sorry, using the medical knowledge available to them to keep their patients alive.

Left: A man recuperating from near drowning, in bed at a receiving house of the Royal Humane Society, after resuscitation by W. Hawes and J.C. Lettsom, 1780s.

Opposite: Resuscitation kit, 1801–50.

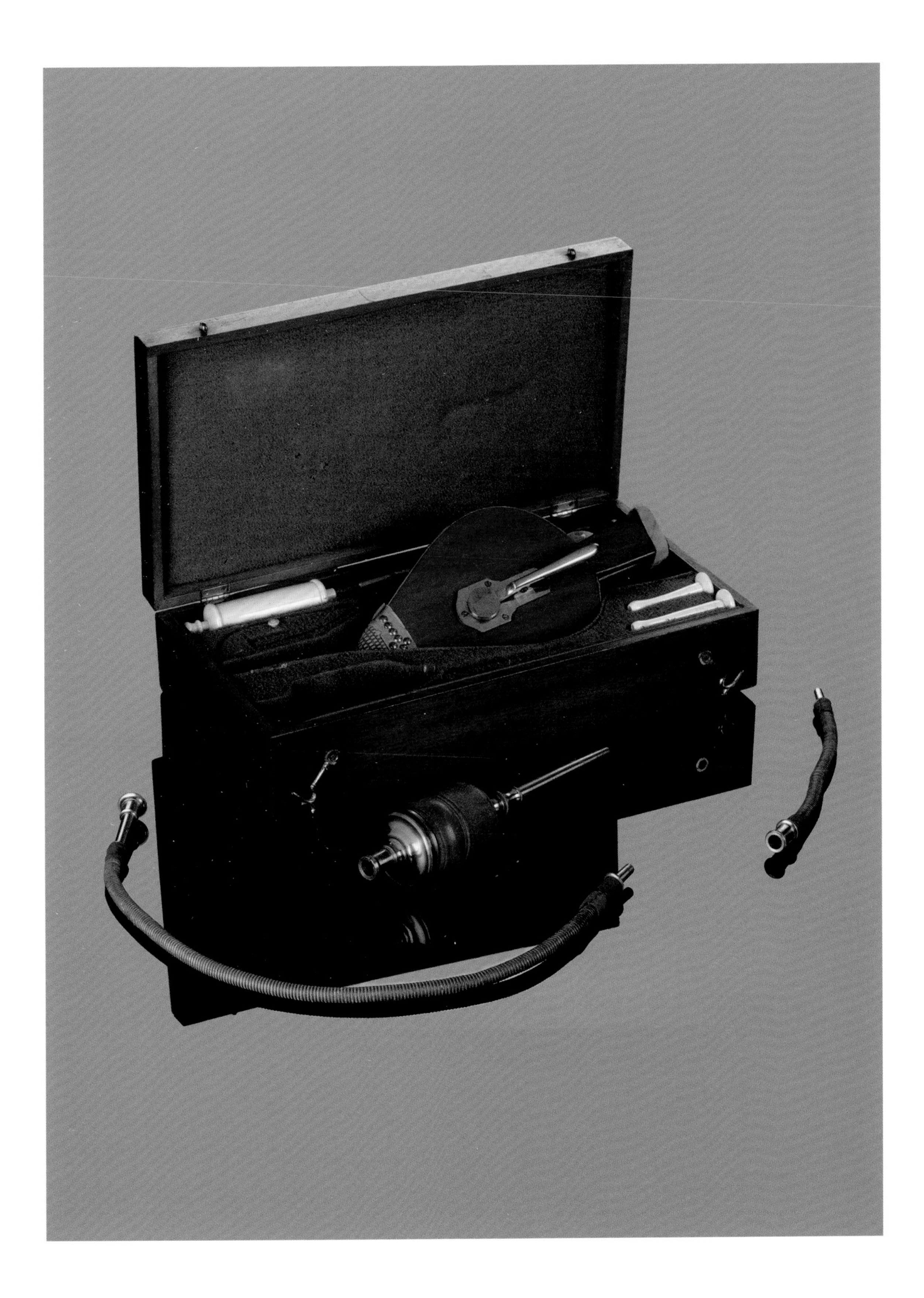

Japanese Bitterlings ♀
For Pregnancy Test.

Ovipositor
Long-
Test
Positive

Ovipositor
Normal
Test
Negative.

Ovipositor
Normal
Partly dissecte

JAPANESE BITTERLINGS

Small Japanese fish called bitterlings react with the urine of a woman to confirm whether or not she is pregnant.

By the 1930s the hunt was on to find a reliable pregnancy test for women. The earliest tests able to detect pregnancy were developed in the 1920s, when scientists discovered that pregnant women produce a hormone known as human chorionic gonadotropin (hCG). They also found that certain animals reacted in predictable ways to this hormone. Mice's ovaries visibly enlarged and matured when administered urine from a pregnant woman. Later tests used rabbits instead of mice. However, carrying out pregnancy tests in this way was far from ideal. Not only did it require hospital laboratories to breed and look after a large number of animals, but in order to see any changes in the anatomy of mice or rabbit, the animal had to be dissected. "The rabbit died" became a common euphemism for a positive pregnancy test by the 1950s.

The hunt was on for an animal model that could show visible changes in its anatomy in response to pregnancy hormones, but that did not have to be dissected in the process. Fish, especially the more translucent varieties such as these Japanese bitterlings, seemed to be a good option. Scientists in the 1930s began to investigate how the females of this species reacted to pregnant women's urine. These three preserved Japanese bitterlings record the results of this process of investigation. The first and the third fish, after being injected with urine from a pregnant woman, had reacted to the hormones by expanding their ovipositor – a tube-like organ for laying eggs. The fish revealed positive pregnancy results, detectable through their translucent skin. The second fish recorded a negative pregnancy result, showing no signs of change. The advantage of this test was that most fish could be examined unharmed and used again in future.

However, the Japanese bitterling test never took off as a reliable means for testing pregnancy. Labs reported that using tropical fish was impractical; the fish were difficult to handle and had a high mortality rate. The test's reliability itself was called into question, as sensitivity to the hormones appeared to fluctuate between individual fish – some gave a positive result even when the urine was from a man. The ideal animal model was eventually found – the Xenopus toad, which reliably reacted to urine by laying eggs, giving results without requiring dissection. All this changed in the 1970s, with the development of modern pregnancy tests that used chemical means, rather than animals, to detect pregnancy.

Opposite: Japanese bitterlings showing positive and negative pregnancy tests, 1930–80.

Above: A positive pregnancy result eliciting a strong reaction from the expectant mother, 1954.

THE EUTHANASIA MACHINE

Philip Nitschke designed this equipment to allow terminally ill patients to end their own lives.

Looking at this curious collection of equipment, it is hard to believe that it once made headlines around the world and still retains the power to provoke questions, be they about the limitations of modern medicine or the sanctity of life. Various items of standard medical and electronic equipment are seen within, and spilling out of, a padded, hard plastic case. These items connect to an out-dated laptop, complete with a floppy disk labelled in pencil with the word "Deliverance". At its sharp end is a hypodermic needle through which, when the machine was activated, a large dose of the powerful sedative Nembutal would pass. In September 1996 cancer patient Bob Dent became the first person in the world to take their own life under a law that allowed voluntary euthanasia – and he used this machine. Three more people did so in the six months that followed.

The man who made the device, which he referred to as the Deliverance Machine, was Australian Dr Philip Nitschke. A former medical practitioner, this author, campaigner and activist is a hugely controversial figure. He created the machine, which incorporated his personal laptop, in response to the law legalizing euthanasia passed by the state parliament of Australia's Northern Territory in 1995. This Act allowed terminally ill patients, having passed a stringent series of approvals and mental and physical tests, to request a medical practitioner to help them to end their own lives. After the four deaths the state law was overturned in 1997 by the Federal Parliament of Australia.

Nitschke designed the machine so that the patient, not the doctor, retained control over the final moments of their life. With the bulk of the machine discreetly placed under the patient's bed, and the needle primed and ready in their arm, the syringe driver that fed in the fatal dose was activated via a simple computer programme. On the laptop screen appeared a series of questions to confirm the person's intent to die, with the option of answering "Yes" or "No" through the tapping of particular keys. The final question was: "If you press this button, you will receive a lethal injection and die in 15 seconds – Do you wish to proceed?" Once confirmed, the machine was triggered and, after a pause of several seconds, the small screen went blank.

Left: Making headlines: Philip Nitschke in his office, 1997.

Opposite: The complete setup of the euthanasia machine, 1995–96.

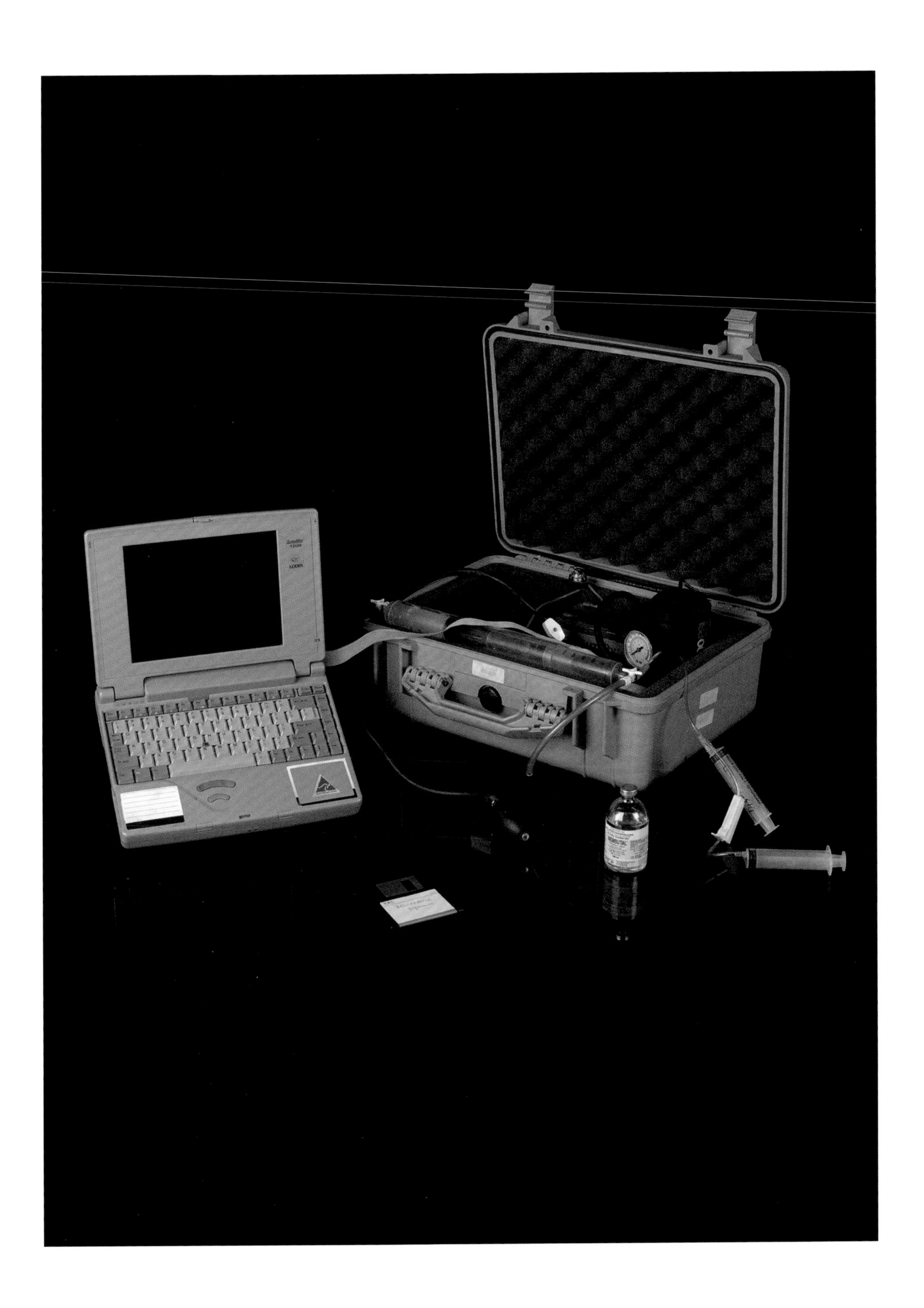
NEMBUTAL

AMBULANCE
Ranger

THE ERANGER – A MATERNAL EMERGENCY SERVICE

This life-saving motorbike and sidecar is specially designed to transport women in labour across difficult terrain.

In 1997, after watching a charity programme about medical transportation in Africa, inventor Mike Norman was inspired to produce a cheaper and more comfortable vehicle. The eRanger is based on a motorbike and sidecar and can be used on a wide variety of terrains at speed. Patients can be transported either lying down or sitting upright in the padded sidecar. A nurse or doctor can ride with the driver if urgent medical assistance is required. In 2017, 99 per cent of all maternal deaths occurred in developing countries, with more than half of these deaths occurring in sub-Saharan Africa, particularly rural areas, due to a lack of access to medical services. The ambulance service is aimed at expectant mothers, or those experiencing complications related to pregnancy and birth.

Developed in Zimbabwe over two years, the first eRanger service was launched in Malawi in 2001. It was reported in the UK House of Commons that, largely thanks to the eRanger, maternal deaths in Malawi dropped by 60 per cent between 2005 and 2009. The service has been adopted by governments and organizations such as UNICEF, the World Health Organization and the United Nations for its programmes on maternal health. Drivers are trained at the eRanger's college in Eastern Cape, South Africa, or in the field, and their training includes motorbike maintenance. The free-of-charge ambulances are contactable by satellite phone and are in use across Africa and the Middle East, with numbers growing year on year. The eRanger has been produced in King William's Town, South Africa, since 2003, providing local employment to the area.

After seeing a news article about the Institute of Mechanical Engineers' 2011 conference, which focused on technological innovations better suited to the needs of people receiving them, curators at the Science Museum approached the developer, Mike Norman. He kindly donated one of the prototypes used to demonstrate the idea around the world. While travelling in an ambulance may be not be comfortable at the best of times, the eRanger has become a lifeline for many mothers and their children.

Opposite: eRanger ambulance, 2008–09.

Above: Training with the eRanger, 2010s.

GANDAU ANCESTRAL FIGURES

Wooden *gandau* models are created to commemorate a loved one's life in north-west Pakistan.

Standing at just over half a metre (20 inches), these figures are known as *gandau*, or ancestral figures, by the Kalasha people who live in three valleys in north-west Pakistan, on the border with Afghanistan. Usually life-sized and carved from a single piece of wood, they are used to commemorate a person after they have died. These two examples show a seated woman wearing a horned headpiece, and a man riding a horse, wearing a headdress. The horned headpiece reflects a belief that a goat born with four horns is a good omen. Male *gandau* figures with headdresses often represent bravery.

When a person dies, the news travels to each of the three valleys and members of the Kalasha community walk to take part in centuries-old customs. A lavish feast is provided for everyone who comes, and three days of dancing and drumming occur around the deceased, with stories being shared about their life. Women from the person's family uncover their own heads, removing the colourful headdresses they wear throughout their lives. The provision of an extravagant farewell is seen as a reflection of a person's social and economic standing. Life-sized wooden figures, intricately painted, are draped with clothing and personal possessions to lead the coffin in procession to burial sites. The *gandau* figures are not purely commemorative but are believed to contain a part of the person's spirit.

The particular figures were purchased at auction in the 1930s, and little is known of their history before this date. However, similar figures exist in other collections such as the British Museum and the Horniman Museum. Curators at these institutions have found that small *gandau* figures were made and sold for the tourist market by the Kalasha people and were collected by British officers during and after the First World War. Their existence at the Science Museum has its roots in the vision of collector and businessman Henry Wellcome, who aimed to collect items reflecting the whole gamut of human experience, including birth and death. Around 10 per cent of his collection is on long-term loan to the Science Museum; the remainder was given to other museums on his death or sold.

A community of 3,500 Kalasha people still live in the three valleys of Pakistan, and have their own language, beliefs and customs. *Gandau* figures are still carved and made but are incredibly expensive, putting them out of the reach of many local people, but they are sold as souvenirs or art pieces.

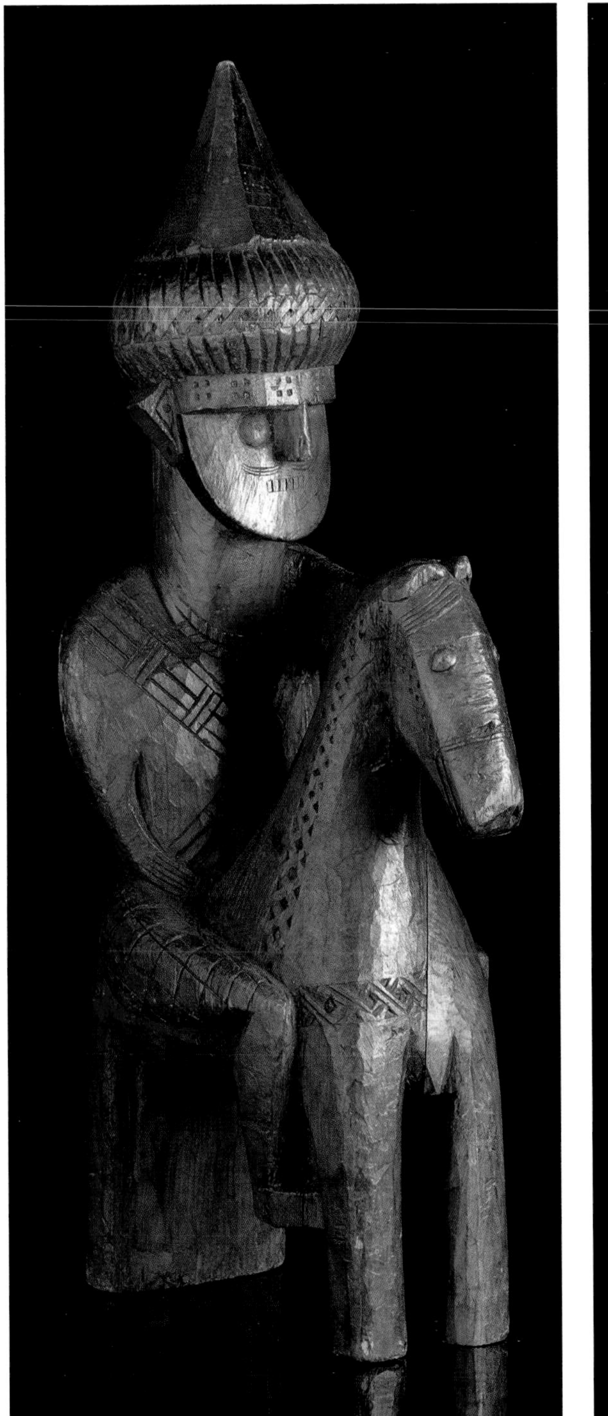

Opposite: *Gandau* figure in a Kalash cemetery, Khyber Pakhtunkhwa, 2008.

Above: Carved wooden model *gandau* figures, Kalasha people, north-west Pakistan, 1801–1900.

TEMPUS – FUGIT
SHAPPER FECIT

SKULL POCKET WATCH

A silver pocket watch advises its owner to spend time wisely by reminding them of their own mortality.

How many times a day do you check the time? Imagine being reminded of your own mortality on each occasion. The owner of this nineteenth-century pocket watch would open the skull to be confronted with three miniature skulls and crossbones, along with the words *Tempus Fugit*, Latin for "time flies". Skulls, hourglasses and time pieces were a common decorative feature appearing on fashionable walking sticks, brooches and rings, all designed to remind the wearer that life is short and time must be spent wisely. These reminders were known as memento mori – Latin for "remember you must die".

To us this seems a morbid thought, but life in nineteenth-century Europe was full of peril – infectious diseases were rife and often deadly, accidents befell workers in the growing industrial towns, and new technologies such as trains were speeding up the pace of life. But could we learn something from our Victorian counterparts and their willingness to confront their mortality? Since the 1840s life expectancy has doubled in the United Kingdom to around age 80, thanks to improvements in living conditions, control over infectious diseases and better healthcare. Death is an extremely difficult and emotive topic often left unspoken, but attitudes are slowly changing.

This reminder of human mortality has already outlived many of its previous owners. The furthest back we can trace this pocket-watch is to Buckingham Palace in the 1930s. Queen Mary, wife of George V, donated the watch to collector Henry Wellcome in the 1930s. Where she got the watch or how long she owned it remains a mystery, but it was not the only item Queen Mary donated. She also gave numerous skull-and-crossbones models to Henry Wellcome as well as scientific instruments to the Science Museum. Wellcome's ambition was to collect material culture and books relating to all aspects of human life, including death. After Wellcome's own death, his collection was dispersed across the globe with the largest portion finding a new home at the Science Museum, including this pocket watch. While the pocket-watch has long stopped ticking, our fascination with these objects hasn't. Looking into the empty eyes of the skull certainly gives pause for thought.

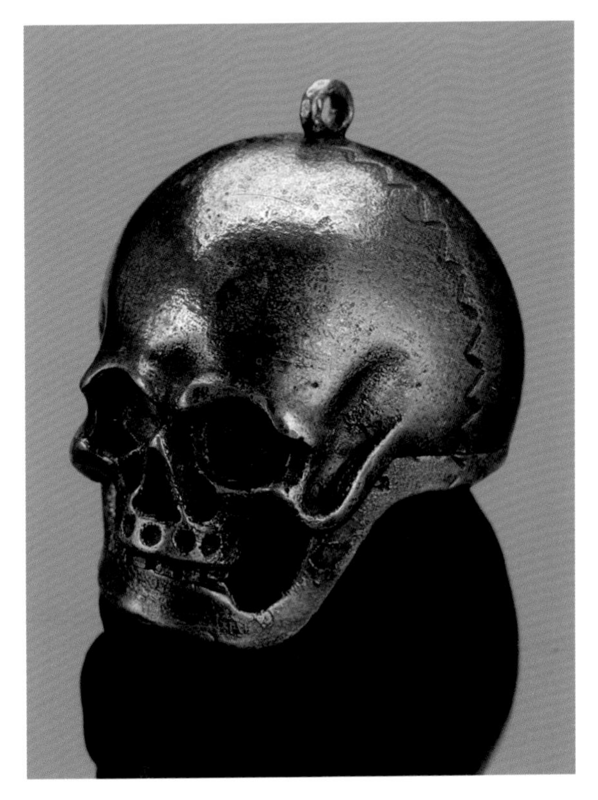

Opposite: Silver pocket watch inscribed with skulls and crossbones inside a skull case, 1700–1900.

Above: The pocket watch closed.

OBSTETRICAL FORCEPS DESIGNED BY WILLIAM SMELLIE

The Chamberlen family invented this instrument to aid difficult births and kept the design a secret for centuries.

This simple pair of forceps is intended to rotate and ease the baby out of the birth canal during a difficult labour. Developed by the Chamberlen family, the design of this instrument was a closely guarded secret from the sixteenth century onward. So clandestine were they that tales abounded of women giving birth while being blindfolded, so they could not report back and perhaps sell the Chamberlens' ideas. Inevitably the secret leaked out and the original basic design was altered by William Smellie. Later known as the "Father of Midwifery", Smellie had originally worked as an apothecary, later studying surgery in Glasgow and midwifery in Paris. He was a talented teacher, and in demonstrations on models and manikins, he shared his method of childbirth with medical students, including showcasing his newly developed forceps.

Supporting a woman in labour is an activity with enormous cultural and personal importance. In most cultures and periods in history, a labouring woman is surrounded only by other women, guided by an older woman with experience of many other local births. Traditionally, this role of midwife has similarities with other forms of healing and medical treatments, carried out by a wise woman with no formal qualifications. As with many forms of medicine and healing, the faith that the patients have in their healer has direct and beneficial health outcomes.

The inclusion of men in the traditionally female space of the labour room was a highly political issue. Female midwives were not considered to be expert enough in human anatomy to wield forceps, and so these instruments were the sole province of the new man-midwives that emerged in the eighteenth century. Not using forceps in a delivery might well have been better for the health of the mother. Despite Smellie's request that the forceps should be covered with removable leather that could be changed, these instruments, often smeared with pork lard as a lubricant, were a perfect way of introducing infection to the reproductive tract. This would often result in the mother's death from childbed fever, which we now know as puerperal infection.

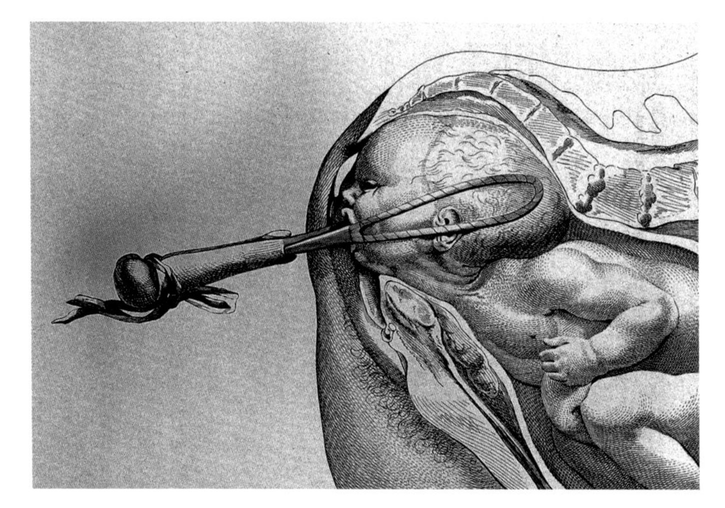

Left: A delivery of a foetus using forceps from William Smellie's *A Sett of Anatomical Tables, with Explanations, and an Abridgement, of the Practice of Midwifery*, 1754.

Opposite: Smellie-type obstetrical forceps, 1701–1800.

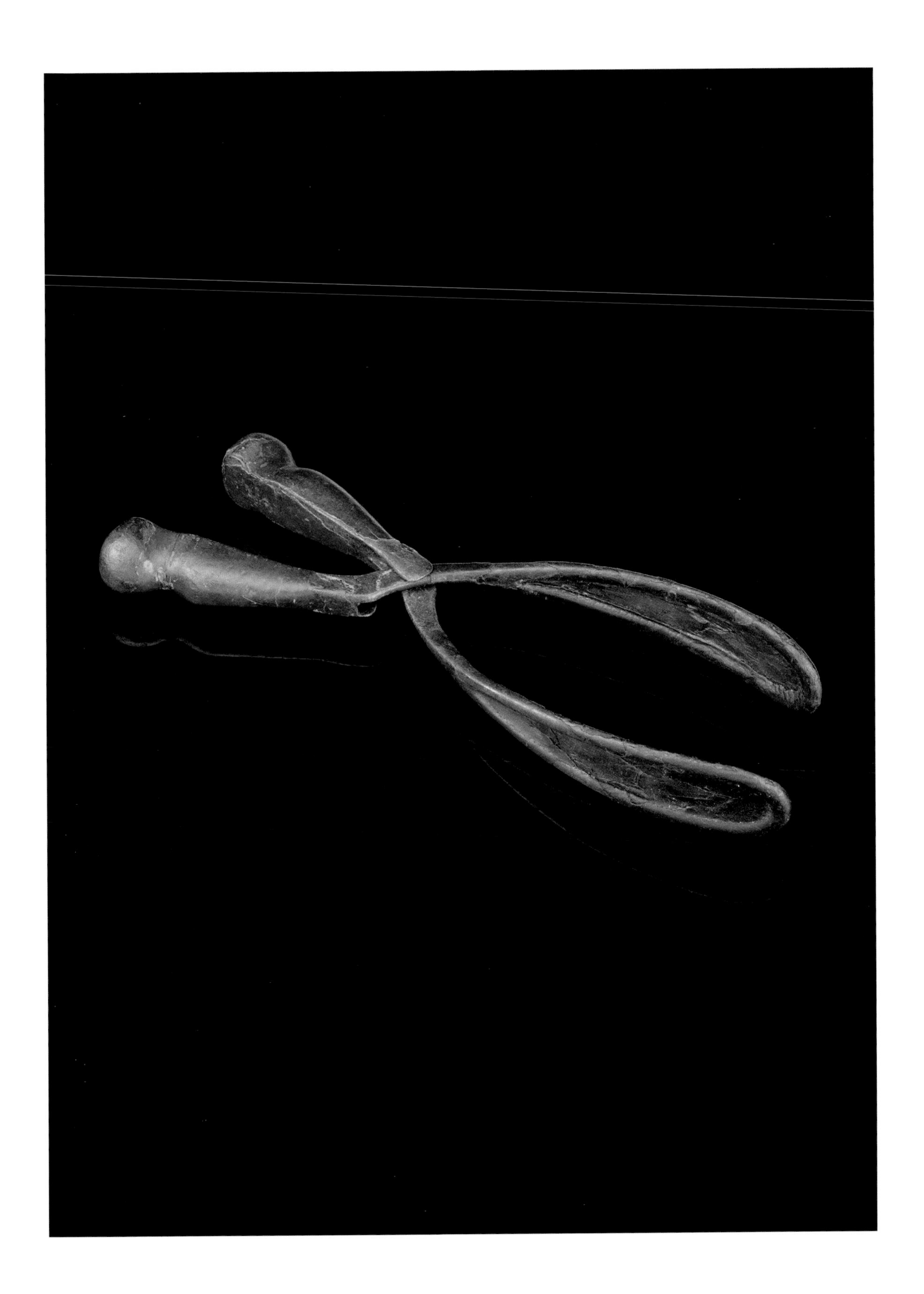

POST-MORTEM TABLE

Best known for creating fine china, Royal Doulton also made ceramic post-mortem tables to be used in the morgue.

A post-mortem, also known as an autopsy, is the examination of a body after death. Carried out by a pathologist, an autopsy is performed to determine the cause of death or to observe the effects and evolution of a disease on the body. The word "autopsy" comes from the Greek *autopsia* meaning "to see for oneself". One of the many specialist tools used is a table, which holds a body at a suitable height, and is cleanable and robust.

To begin with, the body is laid down on a table and washed. The pathologist then performs a visual external examination looking for bruises or any other relevant marks. After this the internal examination can begin. A Y-shaped incision is made with a large scalpel, from shoulder to shoulder and moving down to the abdomen. Rib shears are used to cut through the ribs and access the organs. Each individual organ is removed, examined and weighed. The colour, appearance and weight of an organ can give crucial clues to the cause of death. For example, cancerous tissues are heavier than healthy ones. Once the whole body has been examined it is sewn closed, ready to be buried or cremated.

Post-mortem tables are made of robust material to support the weight of the bodies and withstand regular use. The one pictured is made of ceramic and has raised edges and slopes, which catch fluids pouring from the body as it is cut open and direct them towards a waste drain. This particular table was used until 1944 at the Rotherhithe mortuary in London.

It was made by Royal Doulton Limited, a company set up in 1815, specializing in salt-glaze and stoneware ceramics, stone jars, bottles and flasks. The company pioneered the sanitary revolution by promoting the general use of stoneware drain pipes and water filters to improve living conditions. Ceramic autopsy tables were certainly more hygienic than the wooden models used before; all fluids could easily be washed away from the ceramic surface, keeping it clean and safe for pathologists. Today, however, ceramic has been replaced by stainless steel, a material widely used in medicine due to its durability, high resistance to corrosion and ability to be sterilized easily. Each design change, from wood to ceramic to stainless steel, made it easier and safer for pathologists to conduct autopsies and identify the cause of death.

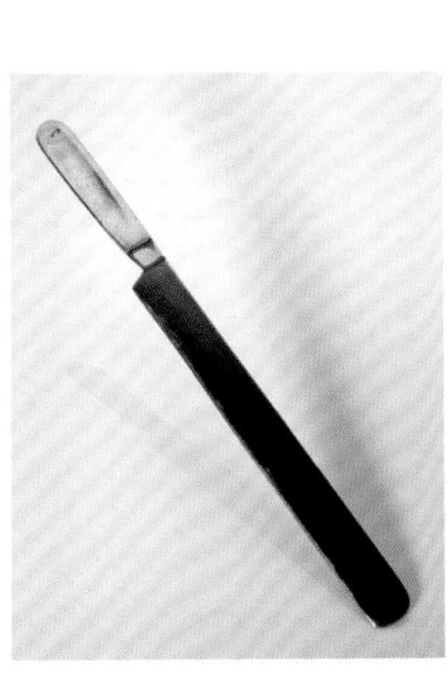

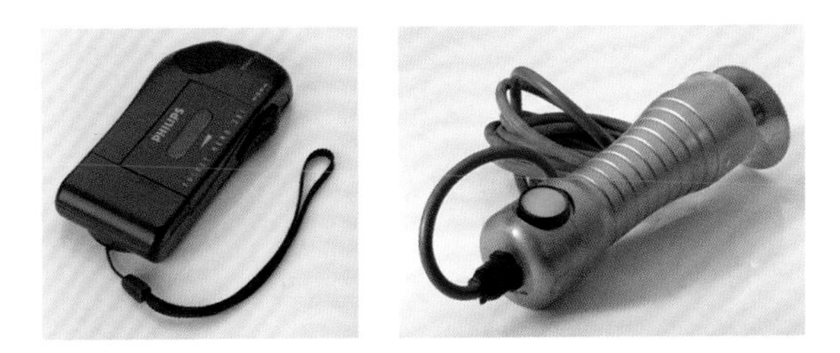

Opposite: Ceramic post-mortem table used at the Rotherhithe mortuary, London until 1944.

Above and Left: Post-mortem tools, 2016.

THE DEATH MASK OF THE KILLER IN THE FOG

Cast from the heads of the hanged, painted plaster death masks were believed to show physical criminal characteristics.

Open the door of a museum storeroom and you might find rows of faces staring back at you. Created during a person's lifetime or, more commonly, after death, plaster masks were produced in an attempt to measure and understand a person's character through their physical appearance, and as a memorial of a loved one. Casts of men of genius, criminals, and people with unusual medical conditions alike were avidly collected by Victorian phrenology enthusiasts, often kept and displayed alongside real human skulls. Phrenologists believed that the shape of the brain can govern the shape of the skull, and that studying both gave a clue to the actions and abilities of the individual.

The method used to create plaster masks is simple. Oil the hair and eyebrows, so the mould can be pulled away easily, and paint a thin layer of plaster over the entire face. A thread is then laid from forehead to chin and more plaster applied on top. Before the plaster is fully set, the thread is pulled to separate the mould into two sections and make it easier to remove. The two sides are then clapped together and filled with plaster to create the mask itself. For a living subject straws are placed in each nostril to allow the person to breathe as the plaster hardens. It is not always easy to tell which masks were taken from a live model and which cast after death, although some faces show evidence of strokes or rigor mortis. A living model might move their face if uncomfortable, and this can sometimes be detected in the final cast.

This death mask shows the Victorian murderer James Bloomfield Rush, known as The Killer in the Fog. Rush killed his landlord's family at Stanfield Manor in Norfolk, and public interest in his crime was so intense that special trains had to be laid on to allow the curious to view the scene of the murders. Despite disguising himself in a wig and false whiskers, he was subsequently identified, convicted and hanged at Norwich Castle. Rush featured in the infamous "Chamber of Horrors" at Madame Tussaud's waxworks in London, which donated this mask to Henry Wellcome in 1936.

Left: *Calves' heads and brains: or a phrenological lecture*, print, London, 1826.

Opposite: Death mask of J.B. Rush, by John Tussaud, London, 1849–1900.

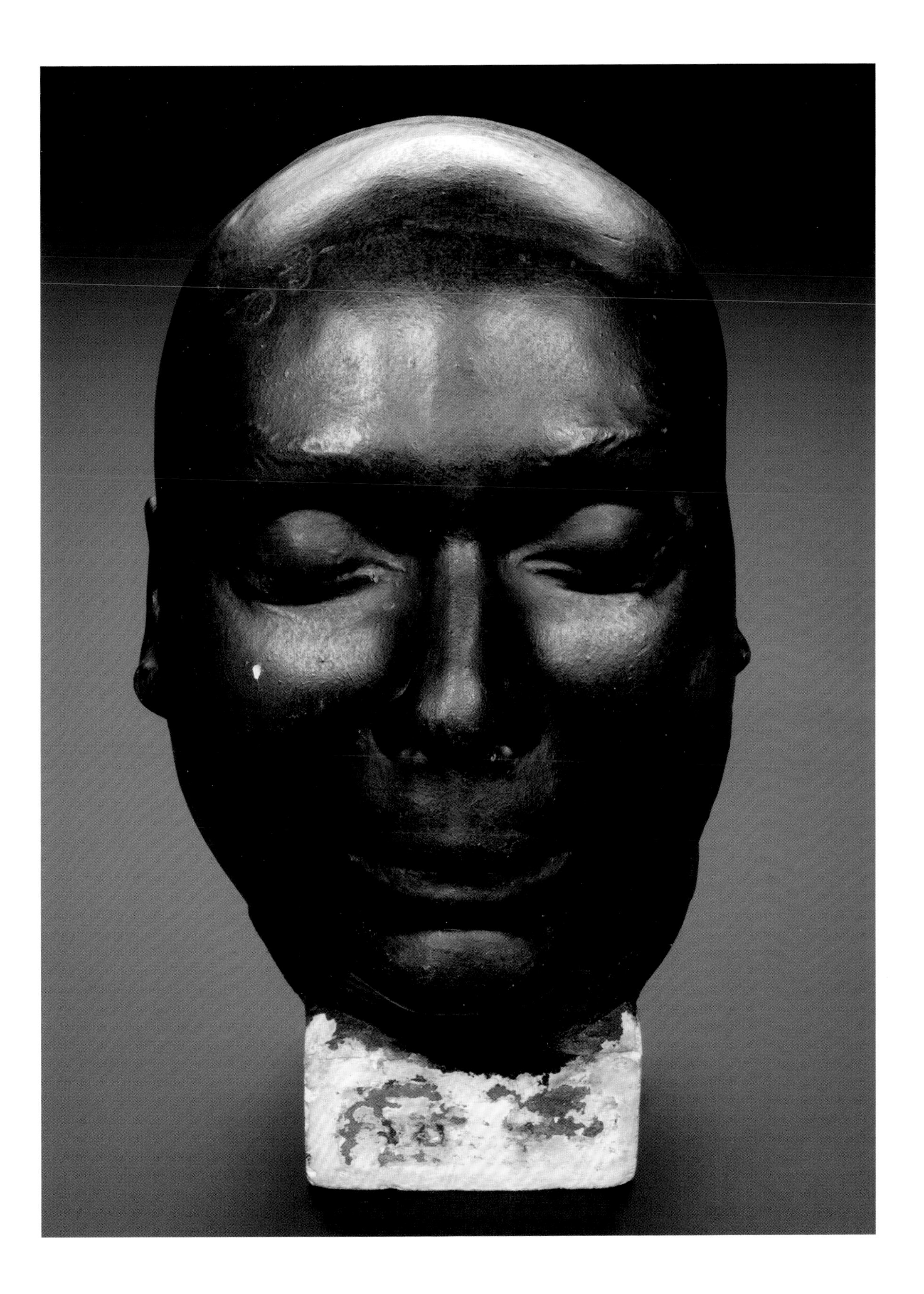

THE LUCY BALDWIN ANALGESIC APPARATUS

Lucy Baldwin campaigned to provide pain relief for women in labour and gave her name to this early "gas and air" machine.

Producing a combination of nitrous oxide and oxygen, this machine was developed specifically to give women pain relief while in labour. Two gas tanks would have been bolted to the side of the machine and set to specific concentrations. The apparatus takes its name from a campaigner for pain relief in childbirth and maternity services – Lucy Baldwin.

Using her position as wife of a senior politician, Lucy Baldwin drew on her own experiences of childbirth, which had been difficult and painful, for her campaigning. In her public speeches she compared childbirth to the experiences of First World War soldiers, saying: "For a woman, giving birth is like going into battle – she never knows, and the doctor never knows, if she will come out alive or not."

Before the arrival of the National Health Service in 1948, access to pain relief while in labour was only for the wealthy. Baldwin was actively involved in the National Birthday Fund, founded in 1928, which concentrated on improving the availability of health services for women who could not afford to access them, on improving young children's nutrition and on providing pain relief in labour. A year later she ran the affiliated Anaesthetics Appeal Fund, to give grants for anaesthesia and anaesthetist posts to hospitals. Baldwin's lobbying for pain relief was met with controversy, even within her own organization. Pain relief in childbirth had been a controversial topic since the expansion of the use of anaesthetics from the 1860s – even despite the advocacy of Queen Victoria, who had herself used chloroform while in labour. The concept of pain was tied up in a complex social, moral and religious rhetoric – and one medical idea was that pain was a useful indication of a patient's condition.

Above: Lucy Baldwin in 1935.

Opposite: The Lucy Baldwin apparatus, 1950–80.

Change came rapidly in Lucy Baldwin's own lifetime, and her campaigning also helped to pass the Midwives Act of 1936, which allowed midwives to administer anaesthesia without the presence of a doctor.

Developed in 1958 and made by the British Oxygen Company, the Lucy Baldwin machine was modified from a dental anaesthetic machine. The ratio of gases was determined by researchers and midwives, who saw first-hand the experiences of women using the machine. This machine was donated to the Science Museum by the obstetrics department of University College Hospital, London, having been superseded by technological developments. Undoubtedly the machine brought welcome relief for thousands of women and carries Lucy Baldwin's legacy of tireless campaigning.

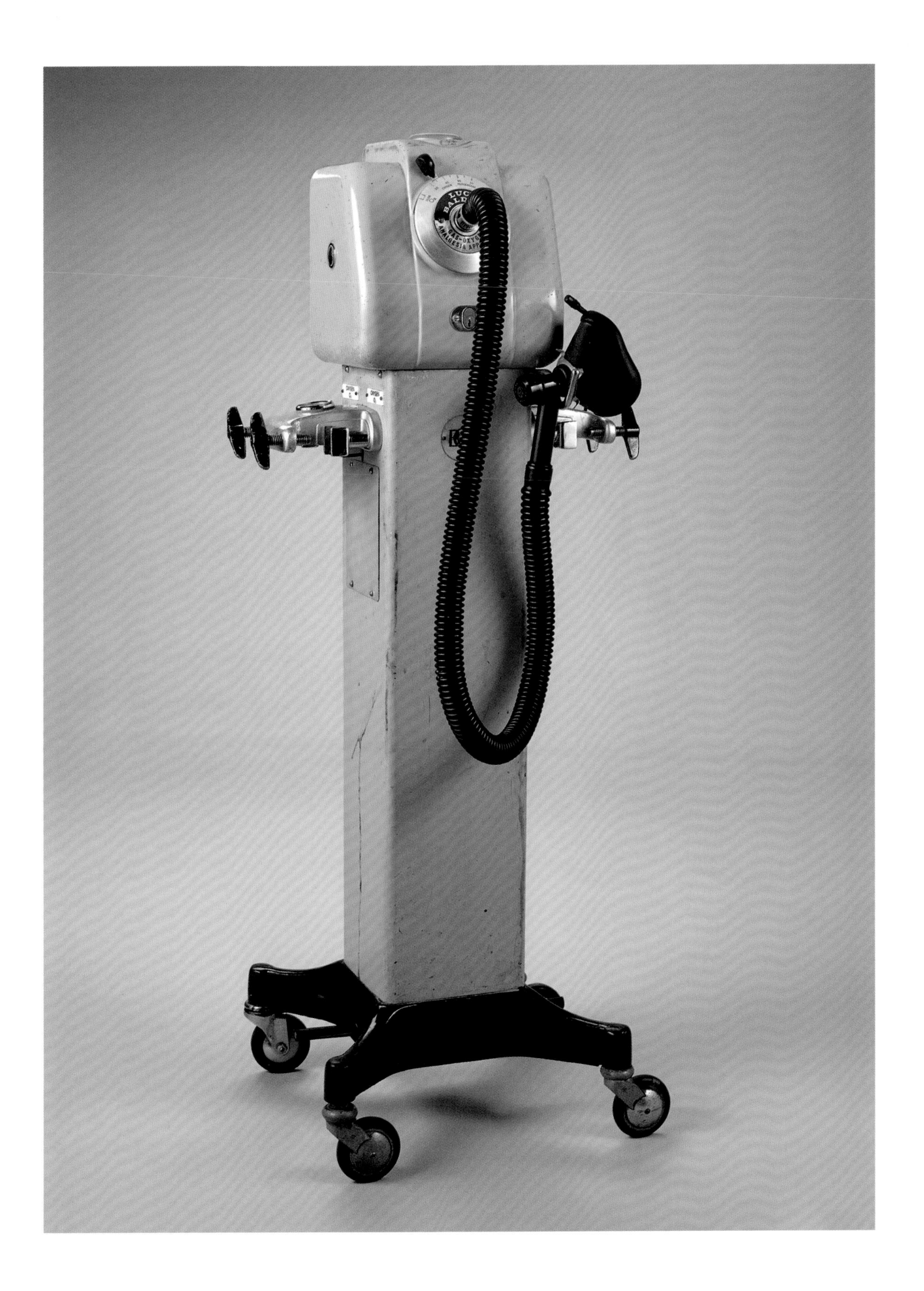

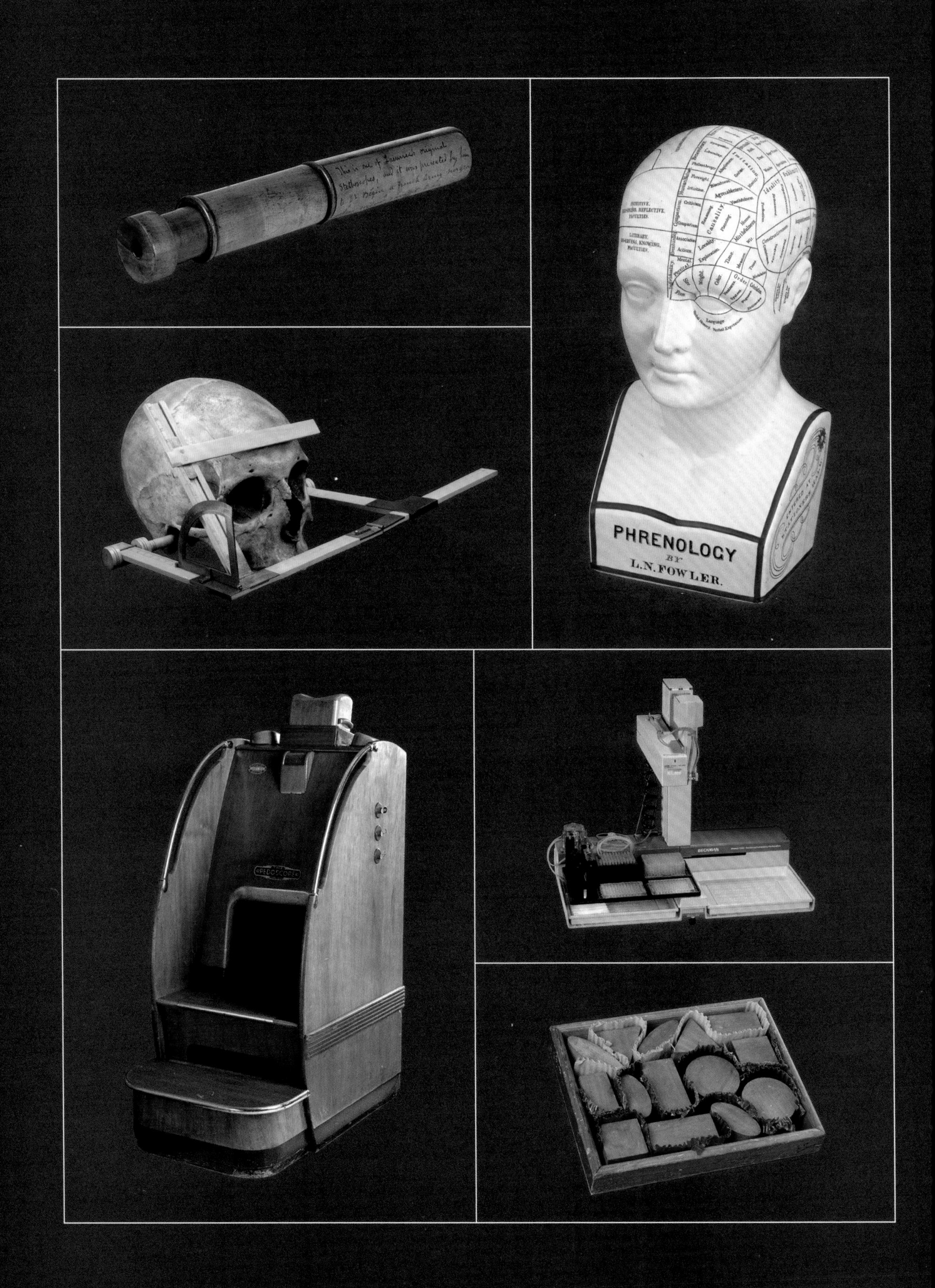
PHRENOLOGY
BY
L.N. FOWLER.
PEDOSCOPE
BECKMAN

3

DIAGNOSIS

When treating an illness, the very first stage is to work out what is wrong. Diagnosis is the detective work carried out by a medical practitioner or an individual, collecting the clues of the symptoms and using experience or research to analyze the condition. A visual or physical examination can reveal the cause of pain or discomfort and many tools are at the medical practitioner's disposal to aid the gaining of information. Testing blood or other bodily fluids, measuring body temperature, or peering into the body's orifices are just some of the procedures that can be carried out painlessly and hygienically with the use of specially designed diagnostic instruments. Sometimes a single symptom can give a clue to the entire problem. Experience is also a crucial tool – a patient may recognize a condition from family history, or a doctor may have seen something similar before.

LAËNNEC'S STETHOSCOPE

René Laënnec invented this special cylinder to allow doctors to listen to their patients' hearts more easily.

Listening to the sounds of the body is one of the most important skills of being a doctor. In 1816 René Laënnec, a French doctor, invented a new tool – a hollow wooden tube – that made listening to the sounds of the body clearer. He named the device a "stethoscope" meaning "to look into the chest". In the two centuries since its invention, the stethoscope has changed in design but it remains one of the most iconic symbols of modern medicine.

For hundreds of years doctors listened for clues about disease by pressing an ear to a patient's body. But this technique posed a number of problems, especially if patients were overweight, infectious or had poor hygiene. Modesty was an issue, especially with female patients. While examining a plump, young woman with a heart condition, to avoid the embarrassment and impropriety of putting his ear to her chest, Laënnec decided to roll up a tube of paper to listen to her heart. He remarked: "I was surprised and elated to be able to hear the beating of her heart with far greater clearness than I ever had with direct application of my ear." Laënnec spent the next three years experimenting and perfecting his design.

As a doctor at the Necker Hospital in Paris, Laënnec used his stethoscope to listen to the lungs and hearts of thousands of ill patients on the wards. He began to match the sounds he heard when patients were alive to the diseases he saw in their bodies when they died. Using dissection, he was able to verify whether his diagnostic theories were correct. Laënnec used lots of imaginative descriptions to categorize the sounds he heard in patients' bodies. A certain type of wheeze he described as the cooing of a wood-pigeon, leading him to remark: "One is tempted to think one of these birds is concealed about the patient's bed."

Laënnec died at the age of 45 from cavitating tuberculosis – a disease Laënnec had helped to identify and understand with his invention. In his own time Laënnec's work was recognized to be a great advancement in knowledge about chest diseases. Preserving his diagnostic devices after his death became important. This stethoscope is labelled: "This is one of Laënnec's original stethoscopes, and it was presented by him to Dr Bégin, a French Army surgeon whose widow gave it to me in 1863."

Left: Laënnec listening to the chest of a patient with tuberculosis at the Necker Hospital, Paris, 1816.

Opposite: Stethoscope made by René Laënnec, *c.*1820.

This is one of Laennec's original
Stethoscopes, and it was presented by him
to Dr Béguin a French Army surgeon
whose widow gave it to me

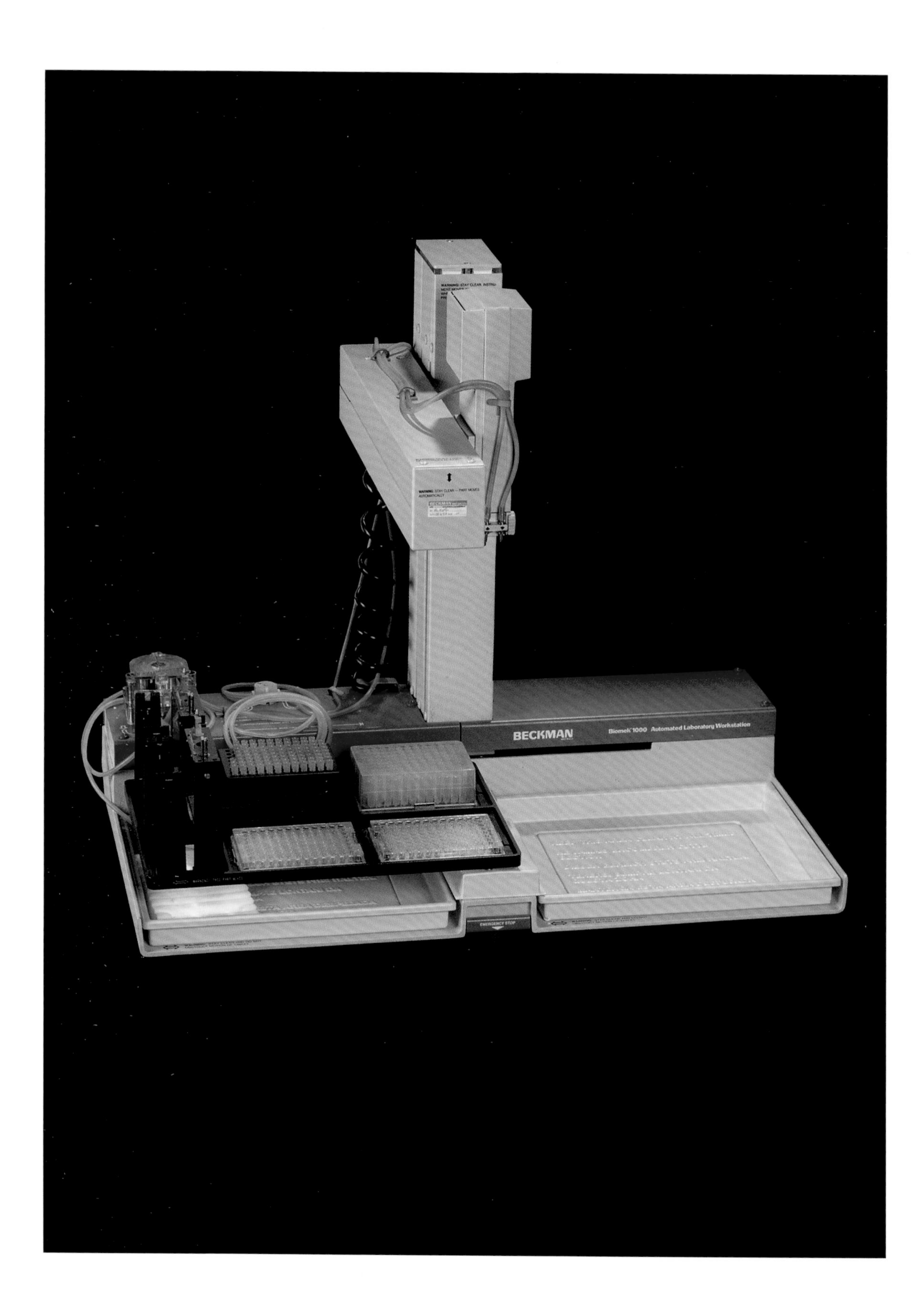
WARNING: STAY CLEAR — PART MOVES AUTOMATICALLY
BECKMAN
Biomek 1000 Automated Laboratory Workstation
EMERGENCY STOP

THE BIOMEK ROBOT

The Biomek 1000 automated laboratory workstation examines multiple biological samples with incredible accuracy.

Whether testing samples of blood, urine, fluids used to test new drugs, sequence genes or diagnose disease, modern medicine depends on moving liquids around fast and precisely to get results. Machines like this Biomek 1000, an automated liquid-handling robot, have become essential apparatus for medical laboratories all over the world. Launched in 1985, one of the earliest laboratory robots, the Biomek 1000 consists of a robotic arm fitted with a tool called a "multi-channel pipette", able to measure and dispense multiple samples of liquids quickly and accurately into uniform plastic 96 "well" trays. Controlled by computer, the robotic arm could be programmed to take samples of liquid from or deliver liquid into the plastic trays, plotting its movements precisely along its vertical and horizontal axes.

Before automated machines like the Biomek were developed, moving biological liquids could only be done by hand. The first tools used to precisely manoeuvre liquids, the burette and pipette, were invented by François Descroizilles, a French chemist and pharmacist, in 1795. But it was developments in the 1970s and 1980s in micro-scale motors and microprocessor technology, that made automated liquid handling robots like the Biomek possible.

Robotic systems like the Biomek transformed the scale and speed of medical research. Our ability to map the three billion base pairs of the human genome was made possible with the Biomek and other similar systems. The ability to process over 8,000 samples a day speeded up the processing of DNA in the Human Genome Project and, just as importantly, reduced human error that could have crept in when having to test thousands of samples by hand. Within hospital laboratories, automation has made the millions of medical tests that are performed each year possible and has reduced the waiting time for patients' results to be returned. One of the most widely used laboratory robots in its day, updated versions of the Biomek continue to be sold to laboratories around the world today.

Opposite: Biomek robot, 1991.

Left: Automated DNA sequencing, 2014.

THE FOWLER BROTHERS' PHRENOLOGICAL HEAD

Phrenological heads showed exactly where the different elements of a person's character were believed to be located.

Think how useful it could be to judge accurately another person's honesty or intelligence merely by looking at them. Although the belief that a person's character can be assessed via their physical appearance is centuries old, the Victorian pseudo-science of phrenology concentrated entirely on the lumps and bumps found on the human skull. Founder of phrenology, German physician Franz Joseph Gall, had studied human brains and, believing that "inner senses" shaped both brain and skull, identified specific areas that would indicate a person's sentimental or intellectual strengths. Developed with his colleague Johann Spurzheim, this new idea was named phrenology, meaning "science of the soul". Many people flocked to get their bumps read and discover if they were especially amorous, secretive or aggressive. Although its followers were keen to be seen as natural scientists, phrenology was widely criticized at the time it was practised. Rather than a social reform movement, it was more in keeping with the characteristics of a belief system and some contemporary literature even referred to it as a sect. Phrenology's appeal was two-fold: it tapped into the Victorian passion for self-improvement and was also very accessible – anyone could do it, given a bit of imagination and a willingness to rummage through someone's hair.

Above: *Bless me what a bump!* (1824–51). A phrenologist analyzes the lumps and bumps on her client's head, with the help of a model head on the table.

Opposite: Fowler's phrenological head, 1860–96. This bust bears the address "Ludgate Circus, London", where Fowler had his own practice from about 1879 to 1896.

Visual charts and models of heads were hallmarks of the practice, clearly indicating the different areas of the skull. Real human skulls, with the phrenological mapping carefully inked on to the bone, were not as popular with the public. Dr Gall himself was not so squeamish; he was known to whip out a real brain from a paper bag during a lecture and give a talk while dissecting it. Charts and ceramic models served as a reference aid for amateur and professional phrenologists alike, and demonstrate a shift away from the individuality of an anatomical model or a scientific specimen towards something more universal. The popularity of phrenology was beginning to ebb when it was given a new lease of life in 1848 with the publication of the Fowler brothers' best-selling book *Hereditary Descent*, followed by a sell-out lecture tour. The Fowlers revitalized phrenology for the remainder of the nineteenth century and their famous stylized head, illustrated here, remains a style icon today.

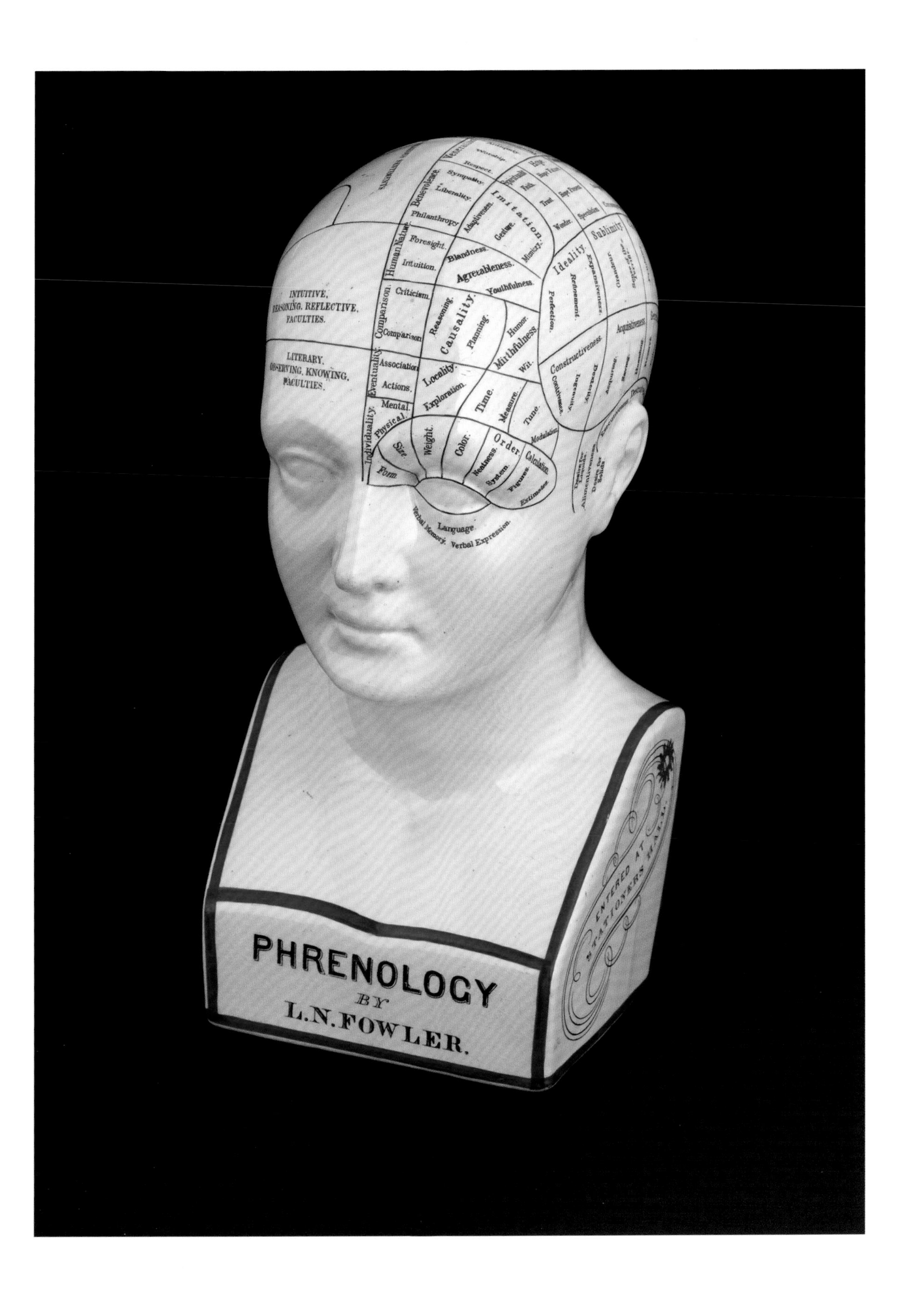
PHRENOLOGY
BY
L.N. FOWLER.
ENTERED AT
STATIONERS HALL.
INTUITIVE,
REASONING, REFLECTIVE,
FACULTIES.
LITERARY,
OBSERVING, KNOWING,
FACULTIES.
Benevolence
Sympathy.
Liberality.
Philanthropy
Respect.
Imitation
Adaptiveness.
Gesture.
Mimicry.
Human Nature
Foresight.
Intuition.
Blandness.
Agreeableness.
Youthfulness.
Ideality.
Refinement.
Perfection.
Expansiveness.
Sublimity.
Wonder.
Trust.
Comparison.
Criticism.
Comparison
Reasoning.
Causality.
Planning.
Humor.
Mirthfulness.
Wit.
Eventuality.
Association
Actions.
Mental.
Physical.
Individuality.
Locality.
Exploration.
Time.
Measure.
Tune.
Modulation.
Constructiveness.
Acquisitiveness.
Size.
Weight.
Color.
Order.
Calculation.
Neatness.
System.
Figures.
Estimates.
Form
Language.
Verbal Memory.
Verbal Expression.

THE
PEDOSCOPE

SHOE SHOP X-RAY MACHINE

Pedoscopes were X-ray apparatus used in shoe shops to check if the shoes were fitted correctly.

Following the discovery of X-rays in 1895, their almost magical ability to reveal the internal structures of the body began to transform late Victorian society. Quickly proving invaluable for medical examination, surgery and cancer therapy, X-rays also infiltrated many other areas of life. Feet-viewing X-ray machines, like this one known as the pedoscope, were an essential fixture of shoe shops between the 1920s and 1960s and became one of the most widespread non-medical uses of X-rays.

Made by the Pedoscope Company in St Albans, the leading British manufacturer of shoe-fluoroscopes, machines like this created X-ray images of the feet inside the shoes being fitted. With the press of a button, you could examine the bones in your feet inside a pair of prospective new shoes. When you consider the anticipation and exhilaration of seeing your own bones wiggling in real time on a green fluoroscopic screen, their popular appeal is unsurprising. During the devices' heyday in the early 1950s, 3,000 machines were in use in the UK and a further 10,000 in the United States.

The basic design of the shoe-fitting fluoroscope consisted of a wooden cabinet containing an X-ray tube in a lead-shielded base. Above it sat a platform, where the customer placed his or her feet. X-rays would pass through the feet and strike a fluorescent screen, revealing the skeletal bones of the feet inside the shoes. The screen could be viewed from three portholes at the top of the device: one for the sales person, one for the customer and another for an observer – usually a mother, due to the machine's appeal among families. This model featured three buttons that delivered different dosages of X-rays and times of exposure for men, women and children.

The era of the pedoscope began to wane from the late 1950s. The dangers of radiation were better known and the public had become savvy to the fact that these devices could not tell you the full story about how well the shoes fitted. A pedoscope failed to show how the fleshy part of the foot sat within a shoe; this still needed to be checked by the store assistant. Concerns about repeated radiation exposure and the health risks the machines posed, predominantly for shoe-shop staff, saw them eventually withdrawn.

Opposite: Shoe-fitting X-ray device known as a pedoscope, 1930–55.

Above: A customer viewing their feet using a pedoscope, *c.*1950s.

CERVICAL SCREENING TEST KIT

George Papanicolaou gave his name to the "Pap smear" test, which checks for pre-cancerous cells of the cervix.

Our understanding of diseases such as cancer has been transformed by a knowledge of how cells work. Being able to recognize the difference between normal and abnormal cells is key to this understanding. The ability to see gradual changes in the shape or behaviour of cells introduced a new era in the prevention, diagnosis and treatment of many medical conditions, including cervical cancer.

While completing research at the New York Women's Hospital in 1927, Greek pathologist George Papanicolaou developed a simple test that used microscope observation of cells from a woman's cervix, the entrance to the womb from the vagina, to diagnose cervical cancer. He identified that cervical cells began to change before they became cancerous. Detecting abnormal cervical cells at an early stage could prevent women from developing cervical cancer. Before Papanicolaou's test cervical cancer was a major cause of death among women in Britain, as the disease could only be detected once it was advanced enough to cause tangible symptoms – by which point the woman's chance of survival is much lower.

In 1964 the National Cervical Cytology Screening Service made the Papanicolaou test, or the "Pap smear" test as it became known, available to self-referring women in Britain who were over the age of 35. In 1988 an improved screening programme was established for women aged 20–64, who were systematically invited for testing every three to five years. The test involved widening the vagina with a speculum, then using a wooden spatula to gently scrape away cells from the cervix. The cells were then "smeared" on to a glass slide for examination under a microscope at a laboratory. Technicians examined manually every cell on each woman's slide (around 80,000 cells per smear test) looking for abnormal cells that could be a sign of cervical cancer. Easy-to-use smear test kits like these contained everything needed to complete a smear test and were used by healthcare professionals throughout the country. This pack was donated to the Science Museum by the London Regional Cytology Training Centre. Founded by Dr Elizabeth Hudson in 1979, the centre provided standardized education and training for those involved in every part of the cervical screening process, from sample-taking to identifying cells. Thousands of these packs have been used, and by 1999 80 per cent of women were being tested regularly and deaths from cervical cancer in Britain had fallen by up to 70 per cent.

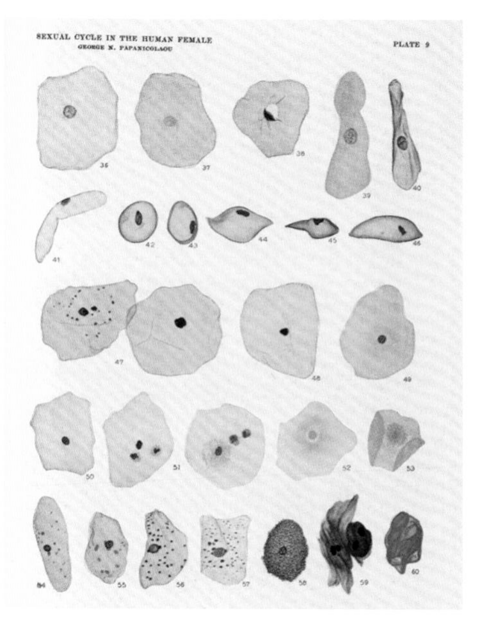

Above: Drawings of various normal cells found in vaginal smears as seen under the microscope, from George Papanicolaou's article *The Sexual Cycle in the Human Female as Revealed by Vaginal Smears*, published in 1933.

Opposite: Kit containing specula, wooden spatulas, glass microscope slides, a pencil and reporting papers, used to complete a "Pap smear" test until 2003.

CerviKit
PULL HERE TO OPEN
CellPath
CellPath
PLACE PENCIL IN HOLE LEAD FIRST
CerviKit™

THE BISCUIT-PACKER'S TEST

This wooden puzzle was used to test non-verbal reasoning skills of workers applying to wrap and pack biscuits in a factory.

If asked to identify a cultural product that sums up Modernism, many people would name the poetry of T.S. Eliot, the architecture of Le Corbusier, the twelve-tone music of Arnold Schoenberg or the artistic creations of Marcel Duchamp. But there is a case also to be made for this curious object, a biscuit-packer's test created by the National Institute of Industrial Psychology (NIIP) in the 1920s. Modernism has been defined by Marshall Berman as the gleeful embrace of modernity, especially of the opportunities that science, technology and new political movements enabled in the first part of the twentieth century, until the Second World War.

Opposite: Biscuit-packers' test, from the National Institute of Industrial Psychology, *c.*1930.

Above: Biscuit packing in a factory, from *How It Is Made*, by Archibald Williams for Thomas Nelson and Sons Ltd, published 1909.

What all the products of Modernism share is a sense that the world can be made anew. One of the underlying assumptions here was that life could be made more rational. That is where the NIIP fits in. Founded in 1919 by Charles Myers, one of the leading figures in the new science of psychology, the Institute was dedicated to organizing work more rationally. One of their key ideas was that low productivity resulted from workers being recruited to jobs for which they were intrinsically badly suited. They therefore developed significant numbers of occupational tests for recruits to different trades, holding that if you were recruited to a job for which you had a positive aptitude, you would not only be more productive in the role but happier, too.

Biscuit packing was at that time not entirely undertaken by specialist machinery; it required a human to do part of the process. In this case there was a job to do wrapping corrugated paper around small piles of biscuits, to keep them separate within an assortment box before the outer wrapper was applied. This was a job that could be done logically, elegantly and fast, or illogically, messily and slowly. How much better for individuals, companies and society, the NIIP reasoned, if the people doing the job were of the former disposition. The test is typical of those areas of work where industrial psychology – like its twin social technology of motion study – had greatest success: namely, women's work in light industry. They had much less success in the more heavily unionized heavy industries dominated by male employees.

PIERRE-PAUL BROCA'S GONIOMETER

Measuring evolutionary development through the facial angles of living patients involved using specially designed tools.

The Victorians had a passion for categorizing and measuring matched only by their enthusiasm for improvement. These two things came together in the science of anthropometry, the measurement of human physical features, which they believed made it possible to calculate laws of heredity. Examining the human body for clues was largely inspired by Charles Darwin's theory of evolution, as outlined in his seminal work *On the Origin of Species* (1859), which demonstrated that all forms of life have evolved and changed over millennia.

Among those inspired by Darwin was the French anatomist and anthropologist Pierre-Paul Broca. Based at the medical school at the University of Paris, Broca founded the Société d'Anthropologie de Paris in 1859. Broca disagreed with some other Evolutionists because he did not believe that all the different human ethnic groups shared a single common origin. His research centred on measuring a wide variety of human heads and faces using a series of specially designed tools. These tools measured facial angles of living subjects, and Broca also weighed and measured the brain size and skull capacity of the dead. The intricate measurements were aimed at calculating how different racial groups had evolved in different ways, through measuring the length of the extension of the jaw and comparing it to the jaws of apes. This goniometer made of brass and boxwood is only one of many instruments used in this way.

The study of physical anthropometry, as practised by Broca and his colleagues, has been called "scientific racism" and with good reason. Subsequent analysis of the nineteenth-century findings shows a clear assumption that white European males were at the top of the evolutionary pile, and this assumption strongly influenced all their findings. They were content to select only the evidence that confirmed their beliefs, changing their criteria as the results emerged. The historian and evolutionary biologist Stephen Jay Gould described Broca's work as: "Heads, I'm superior; tails, you're inferior." In a strange quirk of fate Pierre-Paul Broca's obsession with brain size ended with his own terminal brain haemorrhage at the age of 56.

Above: Portrait of Pierre-Paul Broca (1824–80).

Opposite: Broca goniometer for determining the "facial angle" and "facial triangle" in anthropometrical studies, invented by Pierre-Paul Broca, 1862–1900.

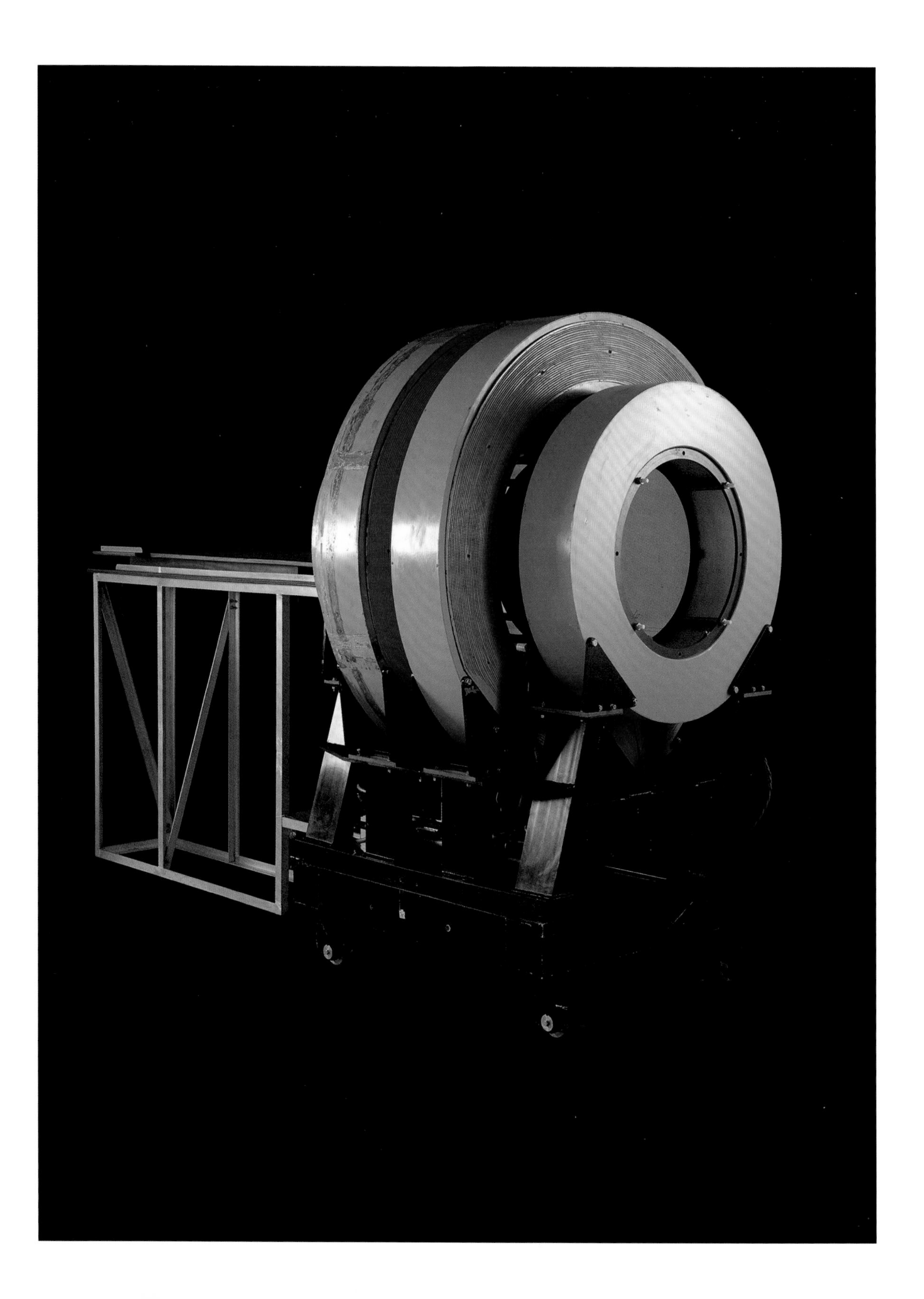

PETER MANSFIELD'S PROTOTYPE MRI SCANNER

This prototype Nuclear Magnetic Resonance machine led to the development of MRI scanning, now used in every hospital.

Magnetic Resonance Imaging, or MRI scanning, is a common medical imaging technique used to look at the body's soft tissues and organs. This cutting-edge technology, ubiquitous in our modern-day hospitals, has its roots in a rudimentary piece of equipment held in the Science Museum's collection.

In the late 1970s physicist Peter Mansfield and his team at the University of Nottingham were working on the development of a whole-body MRI scanner. They had already produced an MRI image of a finger in 1976 and wanted to push their experiment to the next level.

Building a machine big enough and powerful enough to scan a whole body was a huge challenge. On Christmas Eve 1977 the team anxiously awaited the delivery of four large magnet coils to complete the equipment they had been working on. The weight of the magnets was such that the suspension of the van driving them to the laboratory broke. For months the team adjusted the magnets and tested whether they could produce an image of the body, by trying to scan cadavers. Soon it was time to find out if the machine could produce an image of a living body.

The experiment was considered risky as the health threats were not fully understood. Peter Mansfield had received a letter from an eminent scientist in America discouraging him from pursuing his experiment, fearing that the machine would kill whoever stepped into it.

After careful consideration, and doing his own calculations, Peter Mansfield assessed the experiment as safe and decided to test it on himself under the gaze of his wife and the rest of his team. The magnet was turned on; Mansfield heard a "click" but felt no pain. The experiment proceeded. Standing still between the overheating magnets, Mansfield was dripping with sweat for 50 minutes before they obtained the image that he wanted, the image that marked the beginning of MRI.

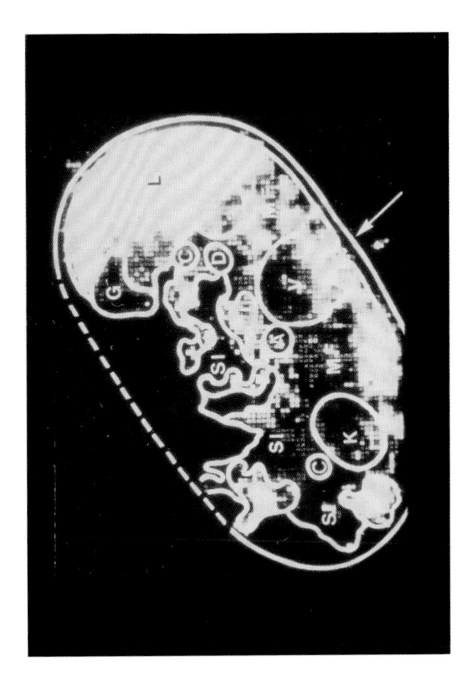

Opposite: Peter Mansfield's prototype scanner, *c.*1978.

Above: One of the earliest images taken using the prototype scanner, 1970s.

MRI revolutionized medicine by providing doctors with detailed images of the patient's internal body structures, making it quicker and easier to examine, diagnose and monitor patients.

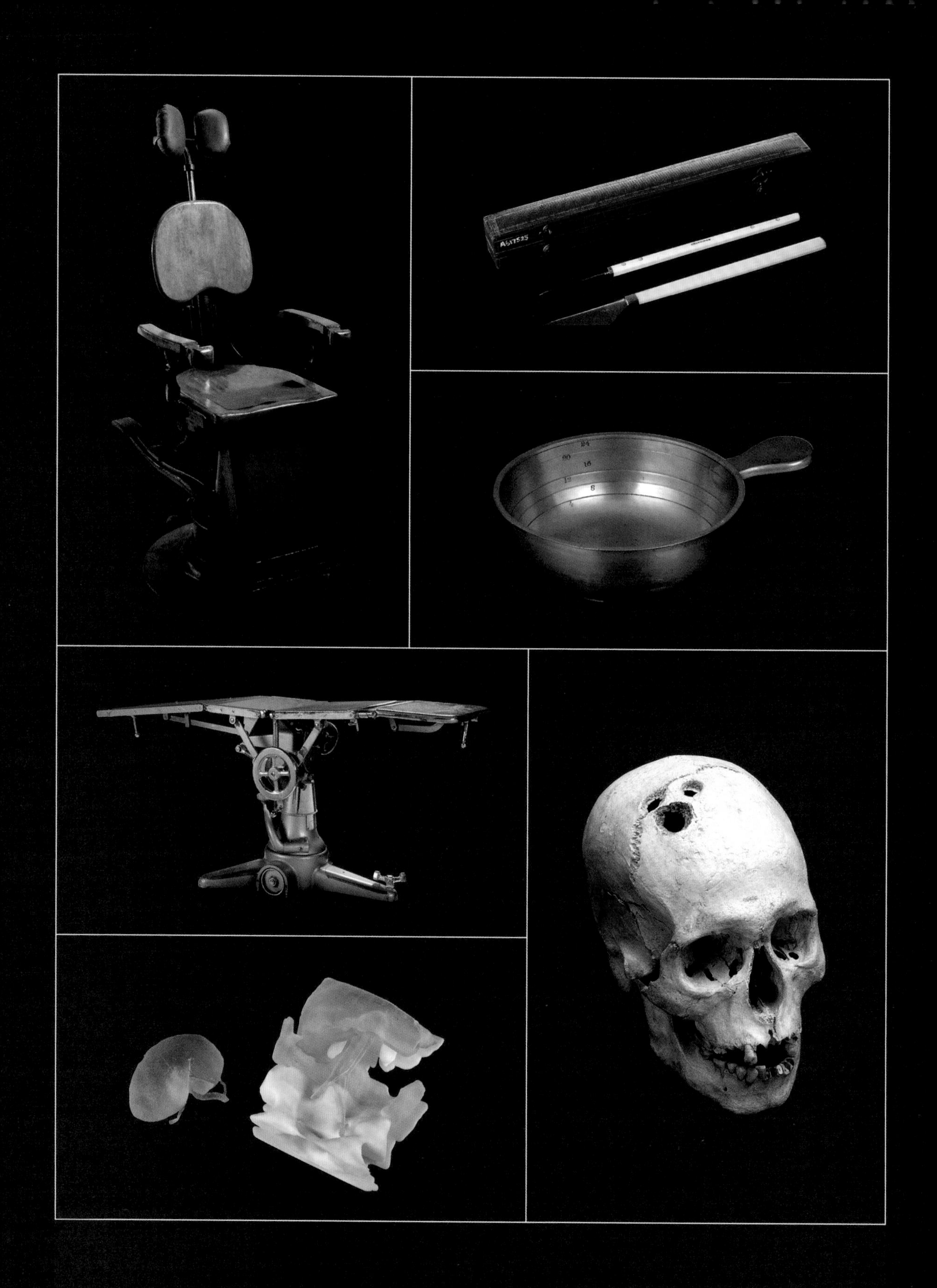
A617525
24
20
16
12
8

4

SURGERY

The surgeon's scalpel cuts through the most complex of all organisms – the human body. Surgical tools, whether an ancient lancet or a recent piece of technology, can tell a story of immense bravery from patients and surgeons alike. Surgical equipment can also create a physical record that traces the development in medical understanding and skill. In the days before anaesthesia or any effective pain relief, only the most desperate would agree to go under the surgeon's knife. Lack of understanding of infection control meant that even if patients survived the operation, they still had a high chance of succumbing to infection even days later. Today, robotic technology performs work with more detail and more steadily than ever imagined in the past, while heart-lung machines blow air across blood to allow complex cardiac surgery to be carried out on living patients.

OPERATING TABLE USED FOR KING GEORGE VI

Used during an operation on George VI in Buckingham Palace in 1951, this adjustable surgical table was the most sophisticated available.

On 23 September 1951 surgeon Sir Clement Price Thomas performed surgery to remove the left lung of King George VI on this operating table at Buckingham Palace, after cancerous tumours had been found. Crowds gathered outside the gates of the palace as a gripped country awaited news of the operation, and the state of the King's health.

Operating tables are one of the most critical items of surgical equipment, essential for situating the patient at the right height and position for the surgical team to perform an operation safely and comfortably. Made by the Genito-Urinary Manufacturing Company Limited, this operating table was one of the most sophisticated models available at the time. It was fully adjustable and made from stainless steel, making it easy to keep sterile. This table came from Westminster Hospital. For reasons of privacy and security, members of the royal family received their medical care at the palace, including major surgery. Hospital staff led by Sister Sarah Minter were tasked with creating an exact replica of an operating theatre at the palace, transferring all essential equipment from the hospital, including this table.

Clement Price Thomas was considered one of the leading thoracic surgeons of his day. He agreed to operate on the King on the condition that he could treat the monarch like any other patient. The surgery was performed successfully. Describing his patient after the operation, Price Thomas said: "The King is the best patient I ever had. Not only is he brave, but he is full of humour. And he is just like an ordinary individual to deal with." Many of the team were named in the King's New Year's Honours List of 1952. After making an initial recovery from the surgery, however, King George VI never fully regained his health. He died in February 1952 at the age of 56. After this, the royal operating table went back to daily use at Westminster Hospital but was fitted with a plaque commemorating its unique history. Though it continued to be used long after King George VI's death, patients might not have known that the table they had been operated on was fit for a king.

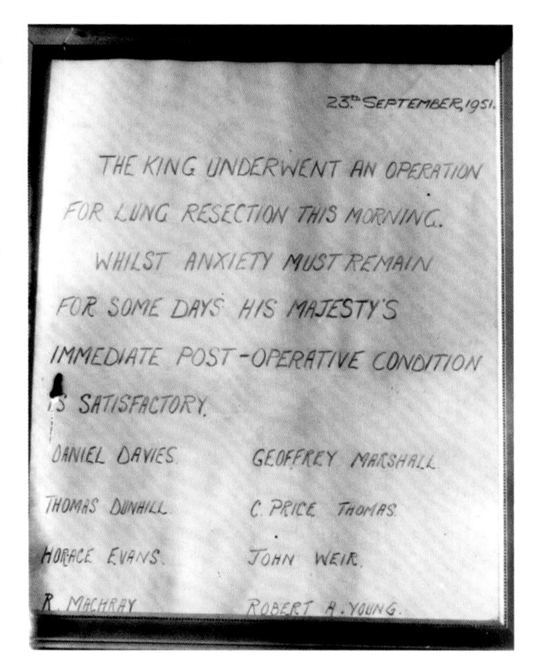

23rd SEPTEMBER, 1951.

THE KING UNDERWENT AN OPERATION FOR LUNG RESECTION THIS MORNING. WHILST ANXIETY MUST REMAIN FOR SOME DAYS HIS MAJESTY'S IMMEDIATE POST-OPERATIVE CONDITION IS SATISFACTORY.

DANIEL DAVIES — GEOFFREY MARSHALL
THOMAS DUNHILL — C. PRICE THOMAS
HORACE EVANS — JOHN WEIR
R. MACHRAY — ROBERT A. YOUNG

Above: 23 September 1951: The second bulletin that was posted outside Buckingham Palace to announce the result of King George VI's operation, which declared His Majesty's condition to be "satisfactory".

Opposite: Adjustable operating table used for the operation of King George VI in September 1951.

24
20
16
12
8
4

BLEEDING BOWL

Drawing blood from a patient, usually into a bleeding bowl, was believed to restore balance to the body.

Feeling weak? Sore head? If you were alive in the nineteenth century, you might have been diagnosed with excess blood. Your treatment required two things: a small, sharp lancet and a bowl. Your doctor would cut into a prominent vein in your forearm or neck and let your blood flow into the bowl until it reached a certain level. The scale engraved on the inside of the bleeding bowl was used to record how much blood was taken.

Today our instincts tell us that bleeding can only be bad for our health and we seek medical interventions to stem the flow when we cut our fingers or graze our knees. So how did bloodletting become one of medicine's most persistent treatments? The practice is thought to have originated in ancient Egypt; it then spread to Greece, where physicians believed that all illnesses stemmed from an overabundance of blood. In the second century CE the celebrated physician Galen expanded and popularized the theory that good health required a perfect balance of four humours – blood, phlegm, yellow bile and black bile. Bloodletting became the answer to an imbalance of blood in the body. Galen's discovery that veins and arteries contain blood and not air, as was previously supposed, sparked the development of a complex system regarding the best time to bleed patients. How much blood was to be removed depended on several factors including the weather and the patient's age. Sessions often stopped when the patient began to swoon, with fainting seen as the natural conclusion of the treatment.

Galen's writings and teachings made bloodletting a common technique throughout the Roman Empire. Before long it flourished in India and the Arab world. In medieval Europe, bloodletting became the standard treatment for many conditions, from plague and smallpox to epilepsy and gout. During the eighteenth century, it became extremely fashionable in Europe and even Marie Antoinette underwent the procedure while giving birth to her first child in 1778.

Opposite: Pewter bleeding bowl, nineteenth century.

Above: A surgeon bleeding the arm of a young woman as she is comforted by another woman, 1784.

The practice of bloodletting remained commonplace into the nineteenth century, but doubts were beginning to creep in. In 1855 Scottish physician Dr John Bennett wrote that he doubted whether bleeding a patient from the arm would do anything except reduce their strength and impede their recovery. By the end of the century new medical theories and therapies came to the fore, reducing the popularity of bloodletting as a way of balancing the humours. Not unlike some of its patients, the practice went out with a faint whimper rather than a bang.

TREPANNED HUMAN SKULL

The ancient skill of drilling holes in the skulls of living people was believed to release evil spirits and cure headaches.

Today, a surgical operation conjures up visions of a sterile operating theatre, gowned and gloved surgeons, and anaesthetists to make us unconscious. However, one of surgery's oldest operations has been carried out for thousands of years: trepanning, or trephining, which means to drill holes into a skull.

In about 2200 BCE the owner of this skull underwent trepanning, perhaps multiple times. Why they had the operation is unclear, but we do know that they survived for some time afterward as some of the holes in their skull show signs of healing. This was against incredible odds, at a time when protection against infection and pain relief was limited to a few plants such as coca leaves. Instruments for the operation were also drawn from the natural world – flint, shark's teeth, shells and obsidian. Skulls from across the globe, from as long ago as 5000 BCE, show evidence of this operation. When one of the first ever trepanned skulls was found in the 1860s, nobody could believe that anyone could survive the operation, particularly as at the time survival of this type of surgery was only 10 per cent – mainly due to the challenges of infection, blood loss and shock. How knowledge of the operation spread, or if ancient civilizations developed it independently, is a mystery. Though no one knows for certain, trepanning was probably performed to help treat headaches and head injuries, for ritual purposes, or to release spirits believed to be the cause of ill health.

In January 1959 archaeologist Kathleen Mary Kenyon presented this skull to Henry Wellcome's growing collection. Her archaeological digs at Jericho between 1952 and 1958 were one of the high points of her career. The skull was found in a tomb in January 1958, in Beth Shan cemetery, Jericho, in the valley of Jordan. It had been buried with two jars and a bronze pin.

Our fascination with trepanning is nothing new. Doctors in the early twentieth century procured human skulls to attempt to replicate the holes and cut marks. In their experiments they estimated that creating a hole in a skull could take as little as 13 minutes. Whether the pain experienced during trepanning was factored into this time calculation is unclear. Though the mysteries of trepanning may never be solved, accessing the brain in neurosurgery today has become routine, controlled and safer.

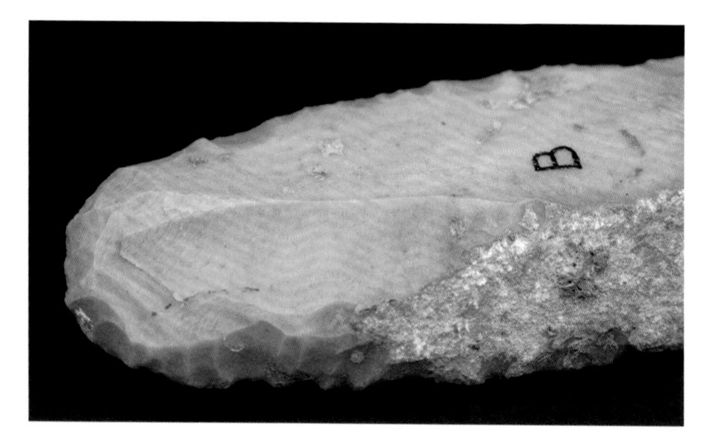

Left: Neolithic flint scraper used for trephination, 2000 BCE.

Opposite: Skull showing four healed trephinings, from Jericho, 2200–2000 BCE.

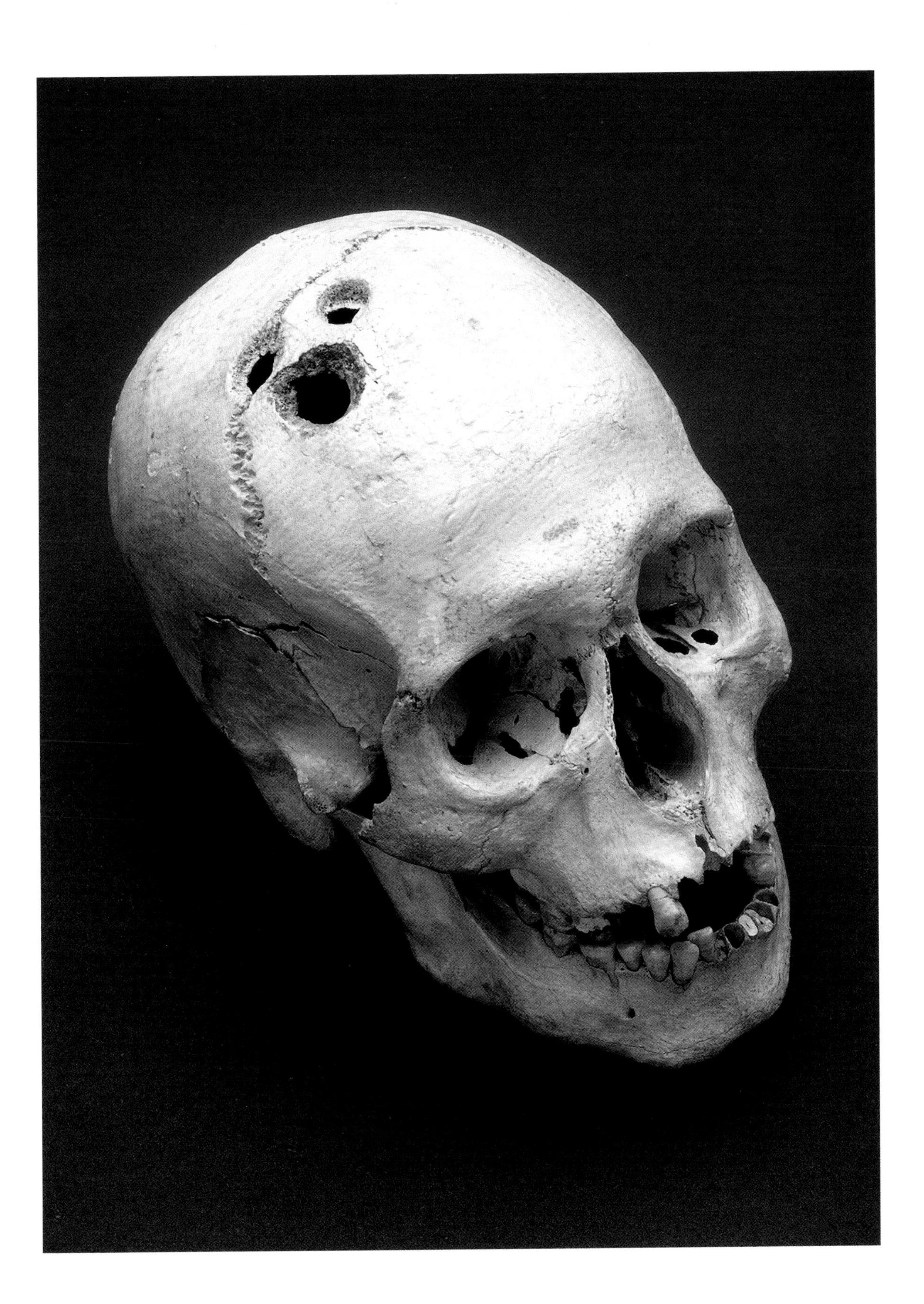

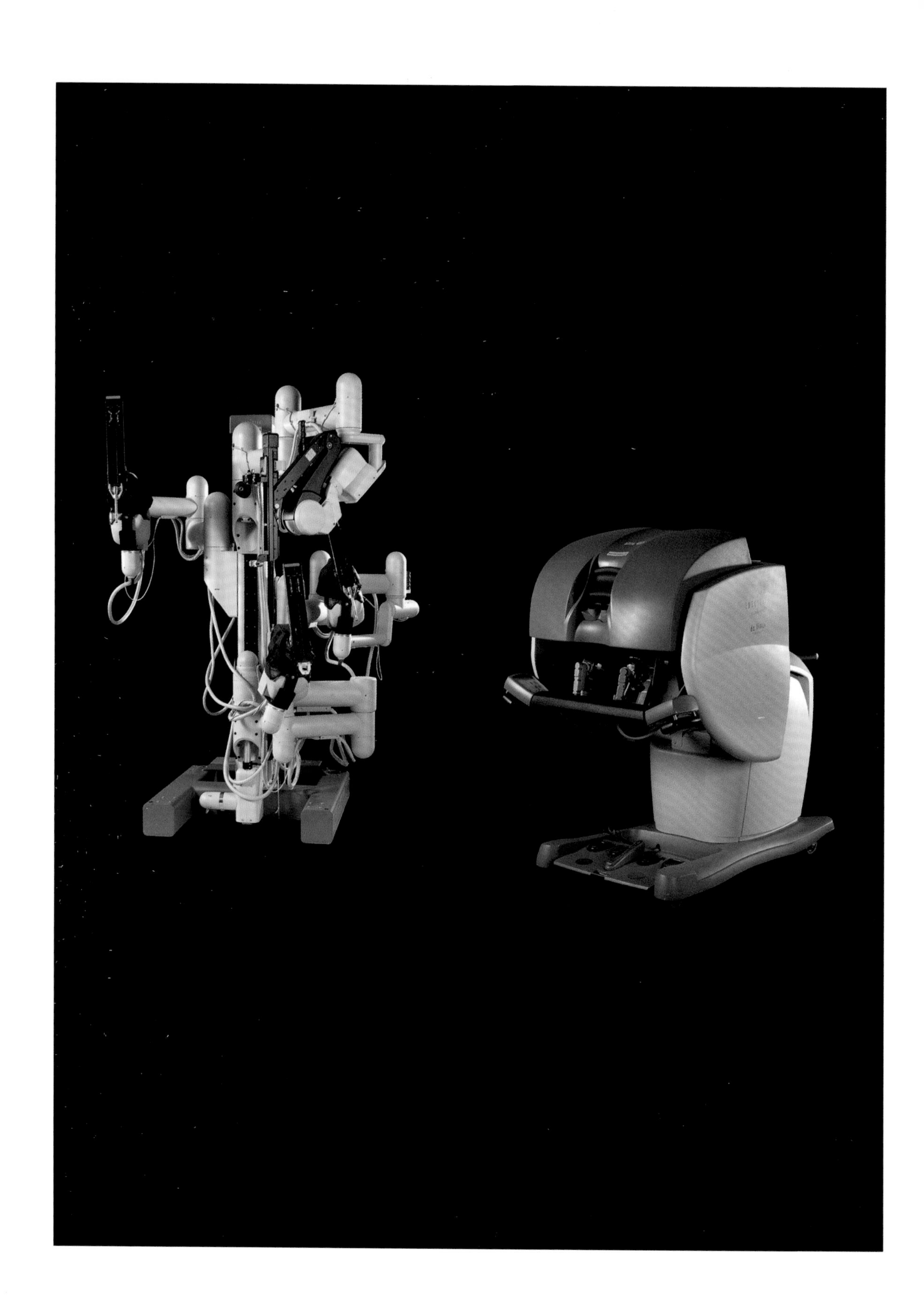

THE DA VINCI SURGICAL SYSTEM

Lord Ara Darzi performed the first robot-assisted surgery in the UK in 2001 using the da Vinci® Classic surgical system.

Imagine your surgeon has come to you to explain your operation and tell you that it will be the first time they will be using a brand new tool – the da Vinci® robotic system. They reassure you that this does not mean the robot is in control, but that they will be operating with the towering two-metre-high robotic arms, and that the da Vinci system is purely an extension of their own hands. This is the scene that played out for one patient at St Mary's Hospital in London in 2001, and this the robot assistant used. It was the first time that the da Vinci system had been used for a clinical case in the United Kingdom.

While operating, the surgeon sits at the console, either in the operating theatre or further away. Looking through two eyeholes, the surgeon sees a 3D image of the inside of the patient's body, provided by an endoscopic camera on one of the arms of the patient cart. The surgeon's hand movements at a console are translated into smaller, precise movements of tiny instruments attached to the arms of the patient cart.

The idea for robot-assisted keyhole surgery developed out of an attempt to bring surgery ever closer to battlefield front lines and to astronauts. Since the system was launched in 1999, by Intuitive Surgical, there have been constant improvements in the technology, and thousands of these machines installed worldwide. The system takes its name from the artist Leonardo da Vinci, inspired by his study of human anatomy and his development of automatons and robots. The manufacturers claim their surgical system improves accuracy, gives 3D views of the body, provides a more comfortable operating position and reduces the natural tremor of a surgeon's hands. However, others argue that this system is expensive, takes time to learn, is large and has little proven benefit over conventional keyhole surgery. Additionally, the robotic arms must be draped in plastic as an infection control measure, adding time to the surgery preparation.

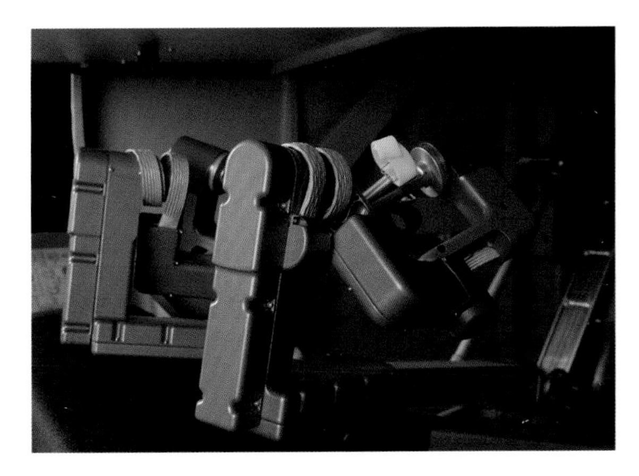

Opposite: The da Vinci surgical system, 1999.

Above: Close-up of the surgeon's consoles showing the hand controls, 1999.

Number 25 off the production line, this da Vinci system has had a varied life, being used in thousands of surgeries and making its movie debut in the James Bond film *Die Another Day* (2002). After its clinical life was over, it became a research tool at the Surgical Innovation Centre, St Mary's, before being donated to the Science Museum by Professor the Lord Darzi of Denham, who in 2001 was its first user.

DENTIST'S CHAIR FOR CHILDREN

Dentists use special chairs to bring their patients to the correct height for examination whilst holding their heads steady.

Dentophobia, the fear of dentists and dental work, is perhaps the most common medical-based anxiety. Such phobias often begin in our childhood. Visits to the unfamiliar environment of a dental surgery, replete with its strange smells, noisy instruments, masked staff and bizarre furniture, can frighten even the bravest of children. At the centre of a visit is the dental chair. The basic wooden-seated dental chair illustrated here, which lacks the cushioned plastic trimming we are now familiar with, is from the early twentieth century. With a simple hydraulic system, patients were raised to an appropriate height for treatment. The size of this example is about three-quarters of a standard chair and designed to seat a specific group of people – children.

While treatment today is free for under-18s, and braces a rite of passage for many teenagers, concerns over the state of British children's teeth continue. Nevertheless, we have certainly come a long way. In 1909 a survey of schoolchildren revealed that nearly 90 per cent of those examined required dental treatment. Such figures added weight to concerns over the link between the dental health of children and the defective teeth being seen in military recruits of the period.

A momentum grew out of these concerns and resulted in the School Dental Service. By the time of the 1909 school survey, a handful of local initiatives had already begun, and eventually expanded across the country. However, progress was not smooth and many children failed to receive treatment. The recruitment and retention of dentists and nurses, to inspect and treat schoolchildren's teeth, also proved to be problematic, and while such care was made a duty of local authorities in 1918, it was only made compulsory in 1944. It was not until the launch of the National Health Service four years later that the situation was gradually transformed nationally. For those lucky enough to receive treatment, these childhood interventions proved to be crucial to the length and quality of future lives; better teeth indicated better general health, physical weight and resistance to disease. It could also mean consistent school attendance, which potentially led to employment. All in all, there was nothing to be afraid of.

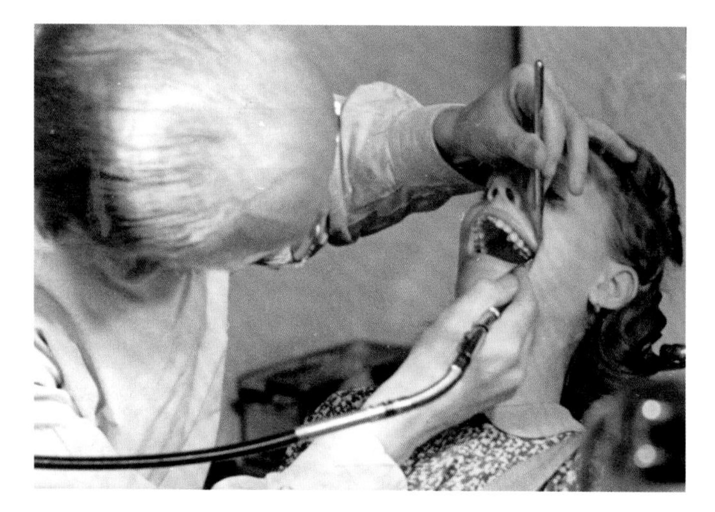

Left: School dentist treating a patient, 1940s.

Opposite: "School" dental chair, by The Dental Manufacturing Co., England 1910–30.

A.43470

LISTER'S CARBOLIC SPRAY

Joseph Lister's invention sprayed surgical patients and equipment with carbolic acid, saving countless people from death by infection.

This odd-looking device is part of the thinking behind one of the most important breakthroughs in our understanding of infection control. Developed by surgeon Joseph Lister in 1860s Glasgow, this device sprayed the patient and medical staff with a fine, pungent, yellow mist of phenol, or five per cent carbolic acid solution. It was part of his system of antisepsis: a method of preventing the growth of micro-organisms responsible for infection – a surgeon's biggest enemy.

At the time surgical death rates could be as high as 50 per cent, making hospitals places to be avoided. Lister believed that the most dangerous infection-causing agents were in the air. He was inspired by the work of French microbiologist Louis Pasteur, who had found that liquids such as beer and milk went bad because of the rapid multiplication of very small organisms – germs. Lister applied Pasteur's germ theory to surgery, exposing wounds to chemical-soaked dressings. In his first published cases in 1867 he reported that the rate of infection was vastly reduced. Lister then experimented with hand-washing, sterilizing instruments and spraying carbolic in the theatre while operating, in order to limit infection.

Made in *c.*1867 by David Marr, a surgical-instrument maker based in Little Queen Street, London, this device featured a spirit lamp to heat the liquid carbolic acid, creating a spray. David Marr worked closely with Lister to develop the device, creating versions that could be used for up to eight hours. Many surgeons followed Lister's practices or adopted aspects of his techniques. Some disagreed with germ theory, while others argued that general cleanliness was improving survival rates. Others argued that Lister's techniques added time to operations, further risking patients' lives. While there had been some knowledge of antiseptics before Lister, he developed a whole system of wound management and surgery.

Lister continually tweaked his practice and theories. Eventually he abandoned the spray in 1887, as he found that germs carried on fingers, dressings and the skin of the patient posed a greater danger. The spray also posed a risk as, for both the patient and the surgical team, inhaling carbolic acid is harmful to the lungs. Although we now associate gloves with infection prevention, or asepsis, the first rubber gloves were used in medicine to combat the irritating effects of the carbolic acid. Added to the many accolades poured upon Lister in his lifetime and after his death, there is still one memento of him to be found in many bathroom cabinets today – Listerine mouthwash.

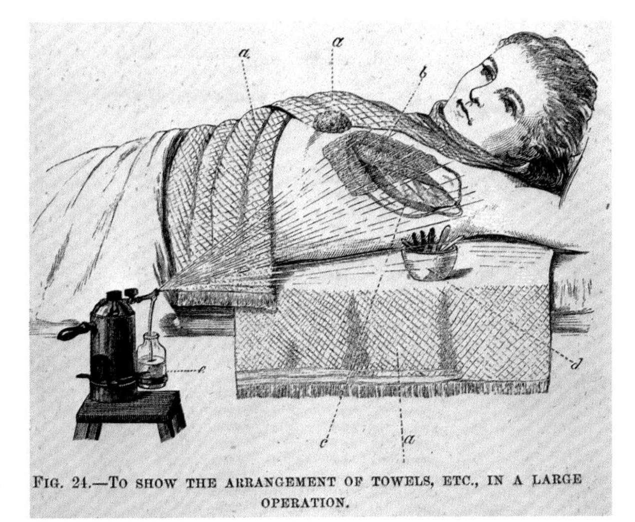

FIG. 24.—TO SHOW THE ARRANGEMENT OF TOWELS, ETC., IN A LARGE OPERATION.

Opposite: Lister's carbolic spray, 1867.

Above: Using a carbolic spray, from *Antiseptic Surgery: Its Principles, Practice, History and Results* (1880) by W. Watson Cheyne.

3D-PRINTED KIDNEY

Planning surgery is helped by 3D scans of organs, such as this complex kidney donation for two-year-old Lucy from her father.

These 3D-printed models are exact replicas of the body parts of two members of the Boucher family – the abdomen of two-year-old Lucy and a kidney from her father Chris. In November 2015 Chris donated his kidney to Lucy; she had experienced heart failure at four weeks old, which caused kidney failure through a lack of oxygen. Until she was old enough to undergo a transplant, Lucy had kidney dialysis three times a week.

Performing a kidney transplant is challenging, particularly when transplanting a full-sized adult kidney into a toddler's abdomen. Planning is essential to avoid any unnecessary delays. While imaging technologies give surgeons a very good idea about the donor's and the recipient's anatomies, sometimes it is not until the surgeon looks inside the body that the full story can be revealed. Pankaj Chandak, a specialist registrar transplant surgeon, came up with the idea of using 3D-printed models after hearing about the use of the technology in paediatric heart surgery. He won funding to develop his idea in August 2015 and – along with colleagues from the transplant team and imaging and medical physics specialists – he developed these models. Lucy's was the first clinical case to use the model and the first in which 3D printing was used to plan a complex adult-to-child kidney transplant.

Creating 3D prints of MRI scans can help surgeons plan an operation in a calm environment and with a wide group of colleagues, creating the opportunity for considering incisions, the approach to the vessels and the best fit for the donated kidney. These models also gave new insights into the tangle of blood vessels in Lucy's abdomen, which were only 1–2 millimetres in diameter. Materials chosen to match the bony, hard pelvis and the much softer structures, such as the liver and the side walls of the abdomen, added texture to the models, enhancing the tactile experience of them. The Boucher family saw the models before their operations, to learn more about the procedure. Chris said: "Seeing the models before Lucy's transplant helped me to understand what would happen and eased my concerns about the surgery. It was reassuring to know that the surgeons could plan the operation in such detail before it took place."

3D printing is a burgeoning field in medicine and holds great promise. Individually tailored 3D-printed implants, medical devices and bone or tissue can already be created, with hopes for 3D-printed organs being available for transplant in the not-too-distant future.

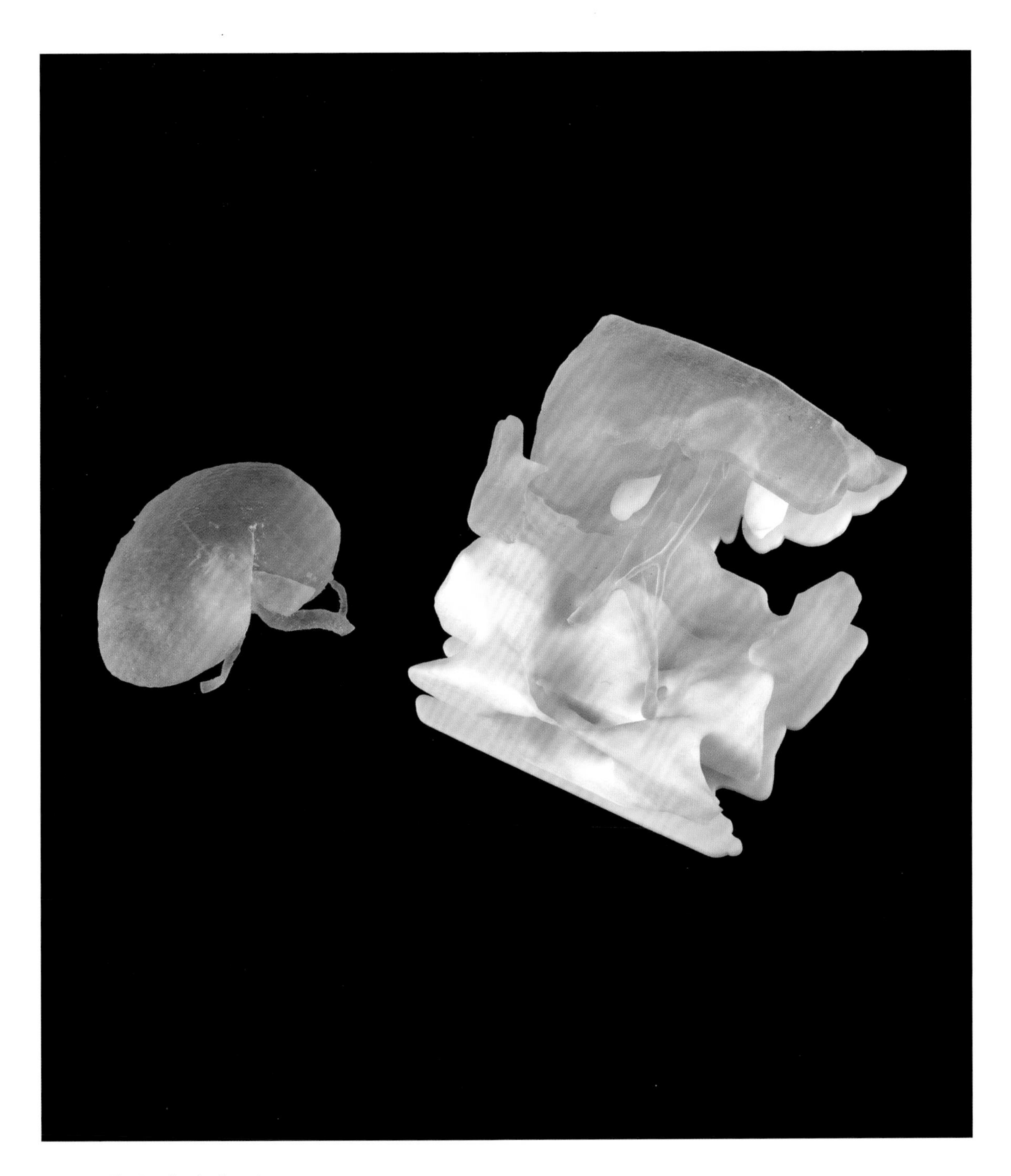

Opposite: The Boucher family and surgeon Pankaj Chandak donating the models to the Science Museum, 2017.

Above: 3D prints of Chris Boucher's kidney and his daughter Lucy's abdomen, before he donated his kidney to her.

A617525

CATARACT KNIFE AND NEEDLE

Removing the clouded lens of a cataract from the human eye requires a steady hand and sharp tools.

Good vision plays a crucial part in how we relate to the world around us. Although progress has been made in the ways people with absent or limited sight are assisted and accommodated, sight impairment can have huge personal impacts. Given the importance of sight, it is unsurprising that eyes, despite their vulnerability and sensitivity, have long been a target for surgical intervention. Historically this was primarily due to cataracts, a common eye condition that occurs when the lens, usually a transparent disc, develops cloudy patches. Over time these grow to cause blurry, misty vision and eventually blindness. Cataracts are associated with old age but can also result from trauma to the eyes or their exposure to radiation and other environmental dangers. Sometimes they are present at birth, but they can also arise from unhealthy lifestyle choices. Cataracts remain a major cause of sight loss across the world.

This cataract knife and needle were tools to treat cataracts. They were originally designed around 1805 by Austrian Georg Joseph Beer, a pioneer of ophthalmology. His technique proved so successful that instruments designed to similar specifications were still being made many decades later. The knife first created an incision into the eye and the needle was then used to remove the cloudy masses and, if necessary, part or all of the lens. In the days before effective anaesthetics, this was an excruciatingly painful process.

Beer was working in a very long tradition, and building on an ancient surgical technique referred to as eye couching. This equally painful treatment for cataracts is mentioned in early texts from India and China, to Africa, Europe and the Middle East – although in the latter, particularly, more sophisticated techniques are also discussed. Couching is still practised in some parts of the world, predominantly where access to modern surgery is limited. Rather than removing the cloudy elements from the lens, once an incision is made a pointed instrument is used to dislodge the lens and push it downward to allow light through. The outcomes from this dangerous technique are limited and, without a natural lens, subsequent vision is inevitably poor – although where strong corrective spectacles are available, the results can be better. Elsewhere, eye surgery now tends to lead to the total removal of the cataract-obscured lens, and its replacement with an artificial version.

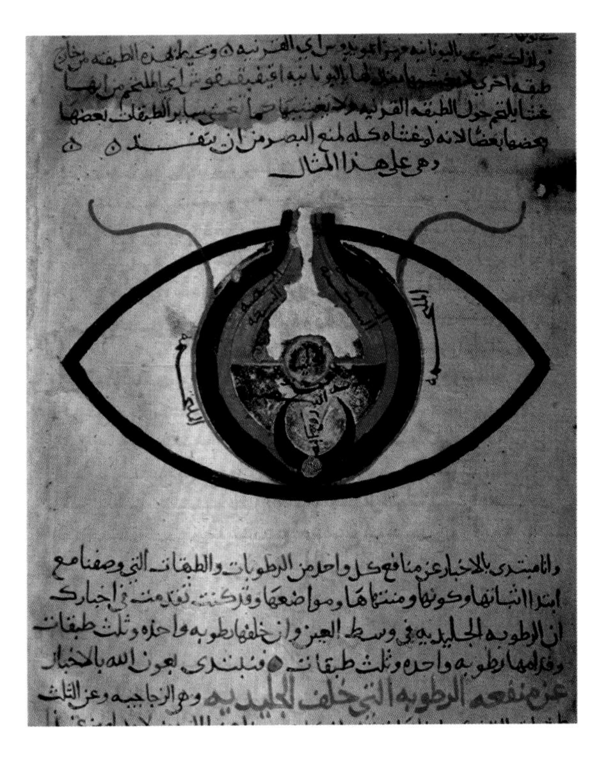

Opposite: Beer's cataract knife and needle, made by Charrière of Paris, late nineteenth century.

Above: *Anatomy of the Eye*, a manuscript by al-Mutadibih dated 1200 CE, is among many ancient texts that discuss cataracts.

HEART-LUNG MACHINE

Heart-lung machines allow for open-heart surgery to take place by keeping the blood flowing and oxygenated during the operation.

This type of heart-lung machine was first used on 17 April 1957, during an operation at Hammersmith Postgraduate Hospital in London to treat a hole in the heart of a 30-year-old woman. A pump takes over the action of the heart, supplying the body with blood. The beating heart can then be stopped, making it easier to operate on. The patient's blood flows over the rotating discs, where oxygen is blown across it, effectively taking over the action of the lungs. This technique is known as bypass. The first heart-lung machine had only been developed four years earlier, when American surgeon John Gibbon developed the technique with Mary Hopkinson, his surgical researcher. Gibbon performed the world's first successful bypass open-heart surgery using his device in 1953, on 18-year-old Cecelia Bavolek, stopping her heart for 26 minutes of the 45-minute operation. However, after three out of his four patients died, Gibbon handed his project over to others to perfect. He did not operate on a human heart again.

Clinical physiologist Denis Melrose was one of the those who took on the problem. Working with Francis Kellerman, founder of medical instrument company New Electronic Products, Melrose developed the first European heart-lung machine. Kellerman took a huge risk on the heart-lung machine, taking out a personal financial loan to cover the costs. The gamble paid off and the Melrose-NEP heart-lung machine, as it became known, began to be used across Europe. By 1968, 1,200 patients at Hammersmith Postgraduate Hospital had been operated on using a Melrose-NEP heart-lung machine. One issue that Melrose solved was a way to safely stop a person's heart. He used a solution of potassium citrate, a technique still practised today, albeit with some modifications. A new specialism grew out of this technology: the role of "perfusionist", whose job is to operate and monitor the machine.

Heart-lung machines make open heart surgery and heart transplants possible. Operating on the heart seems routine today, but at its beginning some saw the heart as a place where surgeons should not meddle, and accused those who did of trying to play God. The heart has strong associations as the seat of emotions and, for some, of the human personality. Revealing the heart to the surgical gaze has changed patients' lives, and our expectations of what medicine can do for us, forever.

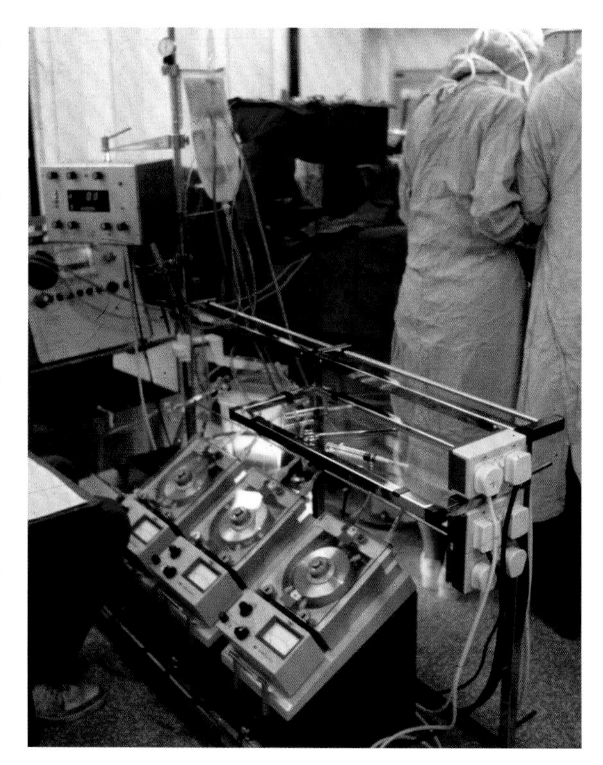

Right: A heart-lung machine in use at St George's Hospital, London, 1979.

Opposite: Heart-lung machine designed by Denis Melrose and New Electronic Products Ltd, *c.*1958.

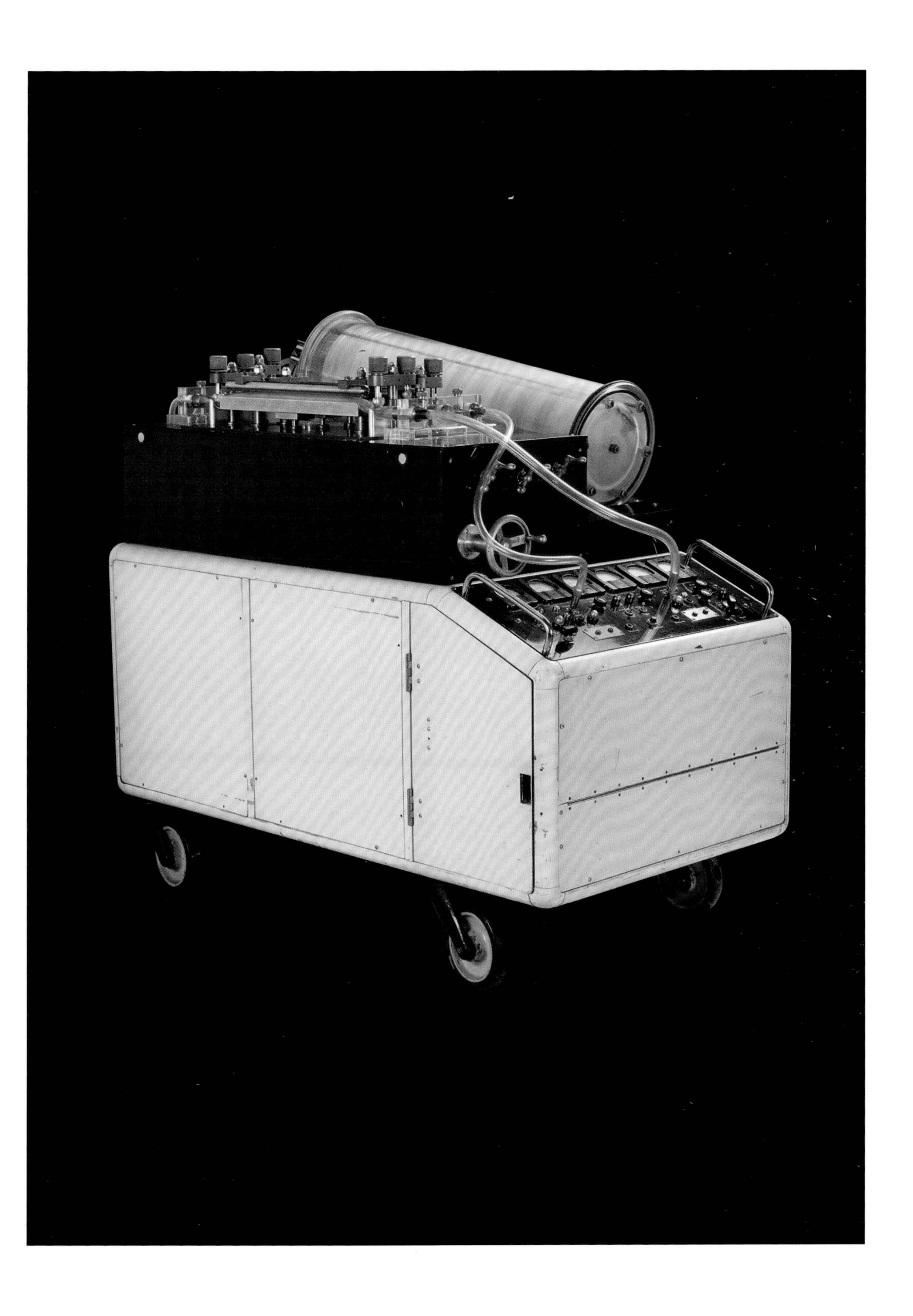

THE BARBER'S POLE

Barber–surgeons advertised both their medical and hairdressing skills with wooden, red-and-white shop signs.

The red-and-white barber's pole is still a familiar sight on the local high street. It is a type of sign used to signify the place or shop where a barber performs their craft. However, the barbers who first used this pole were rather different from the ones on modern high streets. They were more akin to surgeons than hairdressers, and for hundreds of years barber–surgeons were among the most commonly visited medical practitioners. They performed a variety of services from lancing abscesses and setting bone fractures to pulling teeth and, of course, cutting hair. The razor was the indispensable tool of their practice.

The design of the barber's pole signified one of the keystones of a barber–surgeon's practice: bloodletting. Cutting a vein to remove blood from the body was a long-practised therapy thought to restore balance to the body during times of illness. With most physicians believing the "cutter's art" was beneath their station, barber–surgeons, who had little or no formal medical training, cashed in on patients needing to lose a few pints of blood.

While little is known about where this particular sign came from, the twisting red and white design of the pole visually represented bloodletting practice and tools. The red reflected the patient's blood flowing, the white the cloths used to clean up after bloodletting. The shape of the pole represented the rod a patient might grip to help their veins bulge, to make them easier to slice open. The finial symbolized the basin used to collect the blood. Unmistakable in the trade it signified, visual signs like this could be easily understood by a public that was largely illiterate.

Symbols like the barber's pole were important and recognizable trappings of the guild, or professional association, called the United Barber–Surgeons Company, that formed in 1540. Over time, as surgery grew in status and skill, a divide grew between individuals who operated and those who mostly practised hair cutting. King George II separated the two professions in 1745. So only expect a strange look from your barber today if you ask them for anything more than a trim!

Opposite: Painted wooden barber–surgeon's sign, possibly English, 1720–1900.

Left: A blue and white barber's pole on an eighteenth-century high street.

fawn
blue white

BARBARA HEPWORTH'S SURGERY SKETCHBOOK

Sculptor Barbara Hepworth asked a surgeon friend if she could sketch in the operating theatre.

Barbara Hepworth is a modern artist whose work needs little introduction. Best known for her abstract sculptures, this sketchbook, however, is a rare treasure from a different period in her work. These are the only surviving sketches from a two-year period that Hepworth spent observing surgical procedures in Exeter and London.

Following her daughter Sarah's osteomyelitis, a type of bone infection, Hepworth met the surgeon Norman Capener, who had operated on Sarah, and he invited the artist to observe operations at Princess Elizabeth Orthopaedic Hospital in Exeter, and later in London. This sketchbook comes from 1948, when Hepworth observed the surgeon Garnett Passe performing fenestration of the ear at the London Clinic. This was a procedure pioneered by Passe (but now obsolete) to correct otosclerosis, a significant cause of hearing loss in middle age, which involved opening a new "window" into the inner ear by removing a small piece of bone. Hepworth was fascinated by the harmony of the operating theatre, and by the parity of Passe's use of the hammer and chisel with her own use in sculpting. She spoke later in a lecture to surgeons in Exeter about how "the privilege of watching surgery meant not only a period of very important study – but it was also an inspiration in complete harmony with what I feel to be the constructive approach to Art as a whole".

Hepworth's sketches of the surgery are rough and hurried, intent on recording the visual relationships and detailed use of instruments that she would need to produce larger paintings. She annotated her sketches with the names of people and instruments, colours, scales, and stages of the operating procedure. Hepworth produced a series of six paintings from these sketches, which are largely titled with the key piece of equipment that they show. "The Hammer", now in the collection of Tate St Ives, is taken directly from the drawing on page 18 (shown opposite).

The sketchbook is inscribed by Hepworth (above) and accompanied by a letter to Garnett Passe's wife Barbara. The Passes remained friends with Hepworth, and Passe spent his last holiday at her studio in St Ives. The private-theatre sister shown in the sketches was Margaret Moir, who later became Hepworth's secretary. Produced in the years when the National Health Service came into being, Hepworth's sketchbook records the kind of medical experience that so many had before a national health service, namely requiring expensive private treatment. She and many other artists became passionate advocates for the Welfare State.

Opposite: Sketch for "The Hammer", from *Fenestration of the Ear* sketchbook by Barbara Hepworth, p.18, 1948, pencil on paper.

Above: Opening page from *Fenestration of the Ear* sketchbook by Barbara Hepworth, p.1, 1948, ink on paper.

Cancer
CHEST X-RAY SERVICE
LEYLAND
OPL 840

5

PUBLIC HEALTH

How best to manage public health will vary depending on where a community is located – an isolated island community will have very different issues from those found in the heart of a city. But clean water, sanitation and a good diet are essential to good health in whatever circumstances people are living. Promoting good health for the entire community can range from preventative measures against infection and the spread of disease to communicating healthy lifestyles through public information campaigns. People have always implemented isolation strategies against diseases such as the plague or leprosy, although quarantine can create additional issues around immunity. Protecting communities through vaccination has been carried out for centuries, and although its efficacy and safety are doubted by some, these concerns have been discredited by medical science. Communicating preventative measures can perhaps play the biggest part in protecting public health – simply encouraging people to wash their hands or give up smoking saves countless lives.

EDWARD JENNER'S LANCET

Vaccination pioneer Edward Jenner used lancets to introduce cowpox into his patients' arms to inoculate them from smallpox.

In the history of medicine, it is very often the innovations in diagnosis and treatment that grab attention and steal much of the limelight. The introduction of penicillin, the rise of new imaging technologies, the first organ transplants are all acknowledged as landmark discoveries. But it is in the prevention of disease and illness, in the actions that reduce the chances of sickness and early death, that some of the most significant advances in medicine have occurred. Such public health measures have often had huge impacts on our wider society. Vaccination is one such advance and, since its introduction more than two centuries ago, millions of premature deaths have been prevented. In the history of vaccination there is no greater figure than the English physician Edward Jenner.

This vaccination lancet, with tortoiseshell handles that can fold to protect the ivory blade, was once owned by Jenner. These small instruments were key both to his initial experimentation and to the early uptake of the vaccine he introduced to protect against smallpox – a disfiguring and deadly disease that was quite probably the single most lethal disease in eighteenth-century Britain. In developing his vaccine, Jenner applied some scientific thinking to common folklore. He was aware of the phenomenon whereby those who had caught the mild disease cowpox were then immune to the related, but far more dangerous, smallpox. He was not alone with this knowledge, with other doctors discussing and even experimenting with smallpox vaccination ideas. But it was Jenner's research and promotion of the vaccine that proved crucial.

Central to Jenner's success was a highly risky experiment: in 1796, using a lancet like this one, he infected an eight-year-old boy, James Phipps. This was achieved by scraping into his arm some pus taken from the blisters on Sarah Nelmes, a milkmaid who had cowpox. Jenner then exposed the boy to smallpox to see if he was immune. Fortunately, his gamble paid off and Phipps did not fall ill. The technique was not without its doubters and, at times, very vociferous opponents – a situation that continues today – but in time vaccination became a standard medical procedure. Today vaccines exist for many diseases, but for a century after Jenner's experiment there was only one for smallpox. In 1980, after a concerted global vaccination campaign, the world was declared smallpox free. This remains the one human disease to have been eradicated through our intervention.

Above: Engraving of Jenner, published shortly after his death in 1823.

Opposite: Jenner's lancet, late 1700s.

EDWIN LANKESTER'S MICROSCOPE

Edwin Lankester examined water from the pump at Broad Street, Soho, the focus of a cholera outbreak in Victorian London.

Produced by renowned London optical instrument makers Smith & Beck, this microscope is important for what we believe was once examined through its eyepiece. It was bought in 1852 by Edwin Lankester, a largely forgotten but fascinating figure in nineteenth-century science. Very much the Victorian polymath, he was a surgeon, naturalist and popularizer of microscopy, but also served as a coroner, was one of London's first Medical Officers of Health and for a time was Superintendent of the Food Collection at the South Kensington Museum – renamed the Victoria and Albert Museum in 1899.

In 1852 one of Lankester's areas of interest was the capital's drinking water. Concerns about what was being drunk by Londoners had been growing for some years. In a report commissioned by one of the many water companies drawing supplies directly from the River Thames, Lankester himself noted that into it "the sewers and drains of a hundred villages and towns are emptied".

The dangers of poor sanitation are no better demonstrated than by the appearance of cholera. In 1853 London became the focus of the third wave of this deadly disease to hit Britain in 20 years. The following summer an intense outbreak arose in the crowded West End streets of Soho, where hundreds died. Lankester chaired a committee of enquiry into the outbreak. Among those reporting to it was another doctor, John Snow. Today Snow is an iconic figure in the history of public health, but back then he was a maverick whose theories that cholera was waterborne, and that the Soho event was due to a contaminated local water pump in Broad Street, were widely derided.

As part of the enquiry Lankester examined the infamous Broad Street water – quite possibly with this very microscope. The committee was persuaded that there was a problem, and after the event the pump was rendered unusable. Lankester recorded that he saw "organic matter" in the sample and considered that the water was likely to be harmful to health. He did not see cholera bacteria, but what he did see reflected the dangerous state of what was on offer to the citizens of Victorian London.

Opposite: The microscope is recorded as being delivered to Lankester on 7 May 1852.

Above: The grim state of water provided to the public is reflected in this illustration by George Pinwell, featured in the magazine *Fun*, August 1866.

SMOKEY SUE

"Smokey Sue", the health education doll, graphically demonstrates the dangerous effects of smoking during pregnancy.

Sometimes you need a practical demonstration to get a message across. This is the case with Smokey Sue, a health education doll that offers a very novel approach to the decades-long war on smoking – a habit that has proved extremely hard to break. Back in 1956 the writing should have been on the wall for smoking. That year British researchers Richard Doll and Austin Bradford Hill published their latest research in the *British Medical Journal*. Building on their previous work, they offered further compelling evidence that smoking tobacco significantly increased the risk of lung cancer. What had long been suspected by some was now clearly proven.

However, by the 1950s smoking was big business, established in the decades after mechanization had allowed for the mass production of cheap cigarettes. By the late 1940s two-thirds of British men and nearly half of women smoked, as did many children. It was extremely popular so, even with such a solid scientific backing, it was many years before large swathes of the medical profession, political classes and the general public really accepted the health dangers associated with smoking. Famously, Richard Doll recalled that when the Department of Health called a press conference in 1957 to share the worrying news, the politician making the announcement was himself smoking a cigarette.

In the years that followed, the same authorities progressed from inaction, through advice and gentle persuasion, to increasingly high levels of taxation and the marginalization of smokers through legislation. Educating the public took a key role in this public health issue and Smokey Sue was a tool to be targeted at particular groups – most obviously older schoolchildren and women of child-bearing age.

On placing a lit cigarette in Sue's mouth and pumping the rubber bulb, smoke was drawn into the water-filled "womb" containing the rubber foetus. The more smoke

Above: Poster from the 1960s, highlighting the dangers of second-hand smoke, especially for young children, was a strategy aimed particularly at parents.

Right: Smokey Sue, ready to be lit up, 1995.

Sue inhales, the more the water discolours and the more tar and nicotine collects at the water line. One of its key messages was to look beyond the smoker's health, to see how their smoke can harm the health of others – specifically the unborn child. Mixing a freckle-faced doll with a realistically proportioned foetus was crude but effective, and very successful. Numerous plastic Smokey Sues continue to smoke for two around the world today.

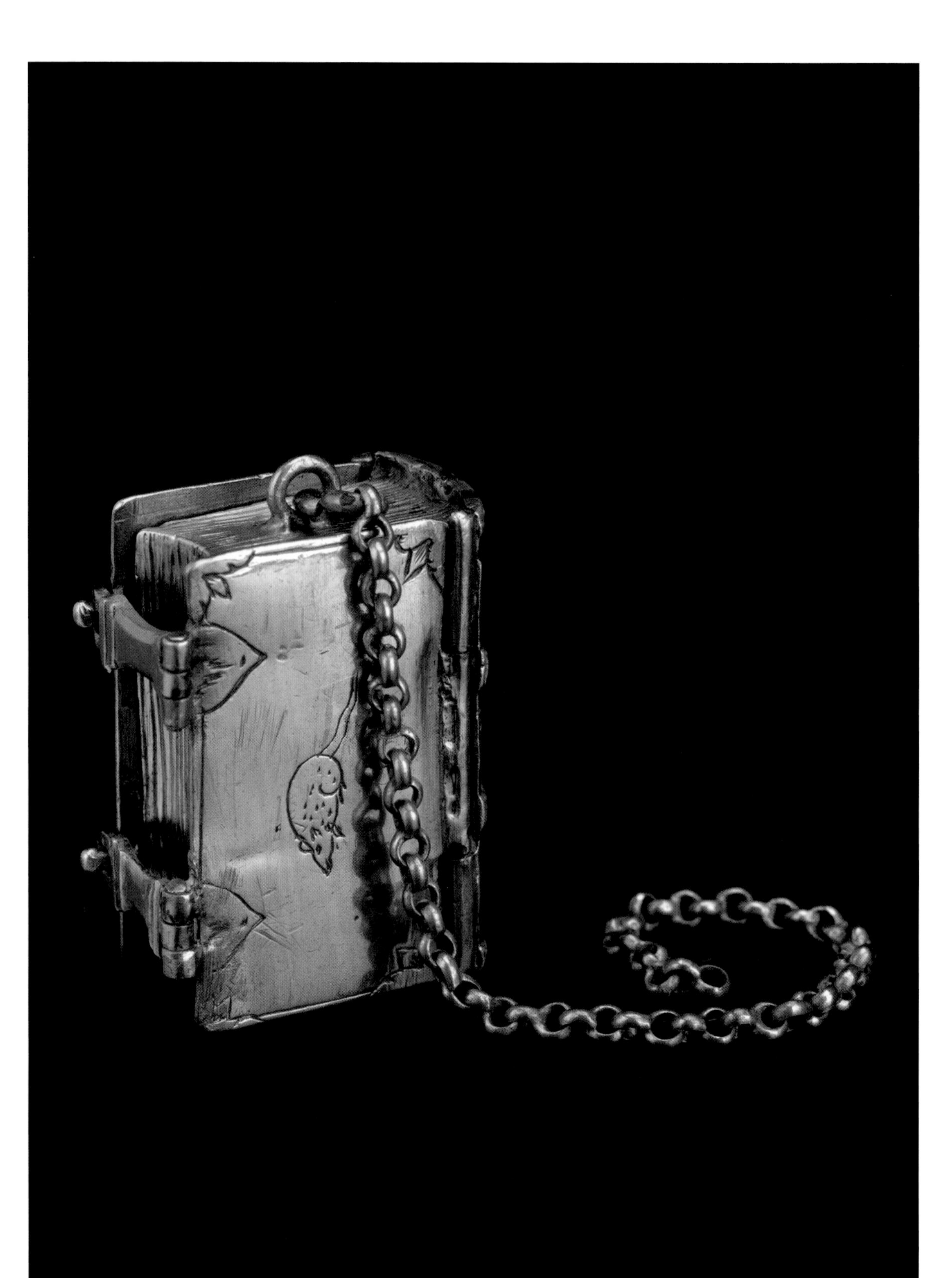

PERFUMED POMANDER

Highly scented herbs and spices in this silver pomander were believed to protect against the dreaded plague.

Shaped like a book, this small silver pomander can be opened to reveal six tiny compartments within. It is on a chain so it could be hung around the owner's neck or about their person, allowing the sweet-smelling perfumes, spices or herbs inside to waft up and permeate the air around the wearer. Pomanders were filled with different types of highly scented material and concoctions thought to prevent diseases spreading through "miasma", or bad air.

The medical theory that miasma – foul or toxic vapours or wind carrying the odours of decaying material – could infect a person or transmit disease was prevalent in Europe from antiquity to the nineteenth century. This phenomenon was also thought to cause many different diseases, including plague. In 1595 petitioners tried to convince Queen Elizabeth I and her government to close a waste disposal site in London and replace it with a garden, as the stench was thought to cause many people in its vicinity to die of plague. Although we now know that *Yersinia pestis*, the bacteria that causes plague, is spread by fleas that feed off rats, this was not known during the Great Plague of London in 1665. There was, nevertheless, a perceived connection between rats and plague as large quantities of rats, just like plague, were thought to be the consequence of miasma spreading across the urban landscape. This can perhaps explain why there is a small engraving of a rat on one side of the pomander.

Wealthy patrons, like the owner of this pomander, might buy ingredients to fill it from an apothecary or perfumer. Books, like *The French Perfumer* by Simon Barbe in 1695, contained recipes so that noble women and less wealthy matrons alike could produce their own pomanders and perfumes in the home. One recipe for "Sweet Bags to wear about you" calls for unspecified perfumed powder, some cloves and some wood to be beaten in a mortar and then sewn up in a four-inch piece of silk fabric. Pomanders and bags like the one described in *The French Perfumer* were used for medical reasons, to ward off miasma and prevent the spread of diseases, but also as fashionable accessories.

Opposite: This pomander would have contained herbs, spices and perfume, believed to protect against plague and other illnesses that spread through miasma or bad air, 1601–1700.

Above: Illustration from *Fasciculus Medicinae* (1495), showing the physician treating a patient afflicted with the plague. He is taking the patient's pulse with one hand and holding a pomander against his face with the other.

THE TRAVELLING X-RAY VAN

Mobile X-ray units drove to city centres and other public places to test large numbers of people for tuberculosis.

British people born between about 1900 and 1950 would be familiar with this public health technology. Mass miniature radiography, started in the Army during the Second World War, was quickly extended to the civilian population afterwards. This mass state intervention in health, started before the National Health Service, focused on a single disease: tuberculosis (TB).

Conventional X-rays had been used in hospitals to diagnose the disease for decades before. But this was different; it is one of the first examples of screening the population for a particular disease. In mass miniature radiography the X-ray image was produced on a glass screen coated with a substance that fluoresced in the presence of X-rays, presenting a temporary image that was then photographed on conventional film using a specially built version of a conventional camera. Eventually the equipment was standardized to use 35mm film, the same as in normal cameras. Using this technique, hundreds of X-rays could be kept on a single film reel. Putting the whole apparatus in a lorry, such as this 1948 "Beaver" van, made it possible to take the chest X-ray service to towns, park in the centre and persuade large numbers of people to be checked. As a punter you would turn up and fill out a card with your details, which would also appear as a cross-link on your X-ray. Once the films were developed, doctors would project them onto a screen, looking for the telltale "shadows" on the lung that suggested the presence of the TB bacterium. If so, you would be invited back to have a full-scale X-ray at your local hospital.

TB in many ways occupied the cultural space that is now taken by cancer. Although caused by a bacterium, identified in 1882 by Robert Koch, this is a case where "soil" is as significant as "seed"; TB is a disease of poverty, flourishing in bad housing where people have poor diets and hard lives. A curious fact about the mass miniature radiography campaigns was that the incidence of TB had long been in decline, largely because of rising living standards – as Thomas McKeon had controversially argued in the 1960s. But before 1952 the main therapy for TB was sanatorium treatments, giving rest and building patients' strength. After this period TB became treatable with the "triple drug" therapy, including the new antibiotic streptomycin. For anyone given a TB diagnosis after a visit to a travelling X-ray lorry like this one, there was, finally, hope.

Above: Weekly schedule for the mass miniature radiography unit around Surrey, England.

Opposite: Mobile mass miniature radiograph unit, housed in a 1948 Leyland "Beaver" van, vehicle registration no. OPL 840.

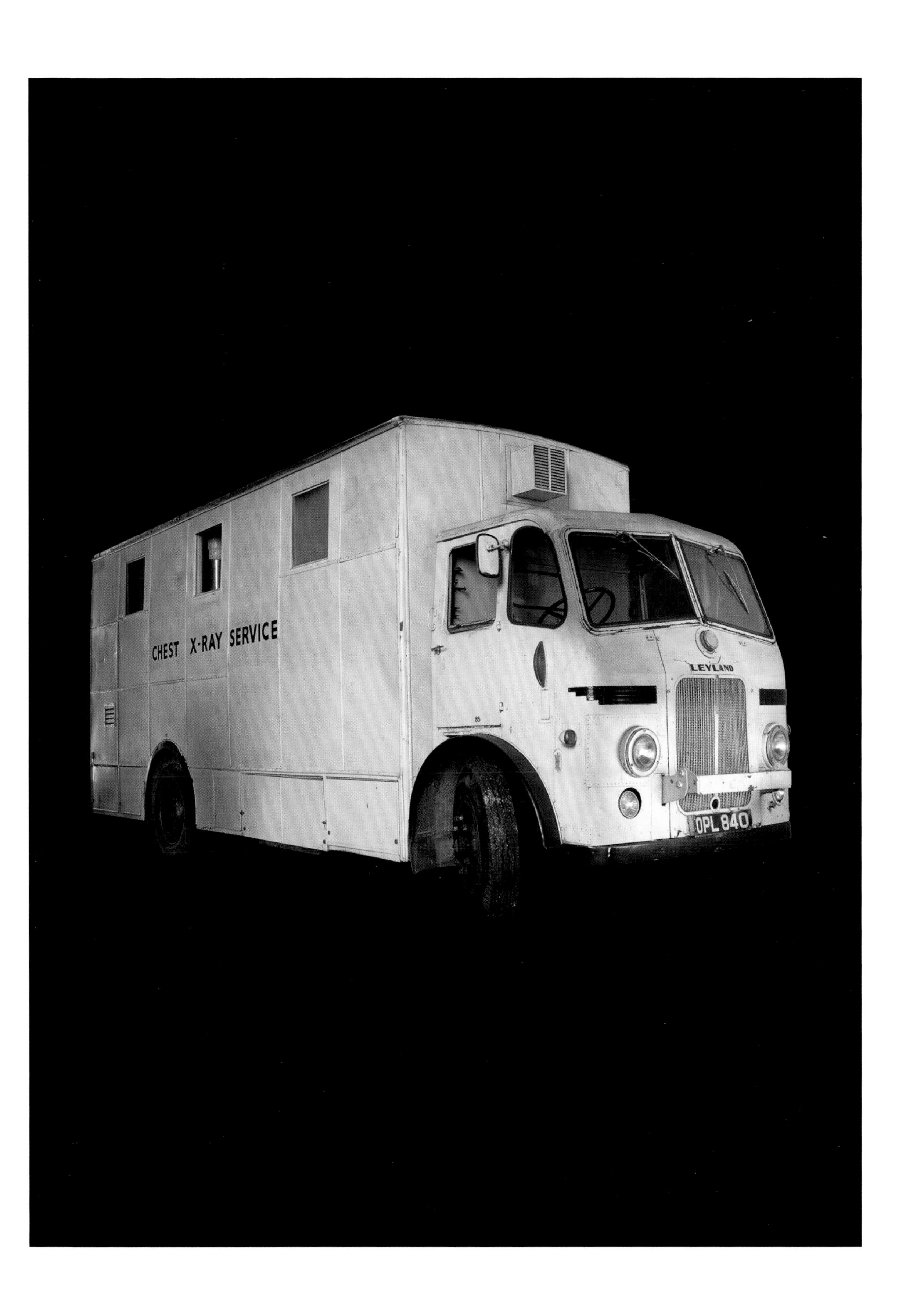
CHEST X-RAY SERVICE
LEYLAND
OPL 840

South Kensington
OCULIS SUBJECTA FIDELIBUS
Lecture Cabinets.
HYGIENE CABINET,
Apparatus and Models of Sanitary Appliances,
For Teaching the General Principles and practically demonstrating Lectures on
HYGIENE AND SANITARY SCIENCE.
DESIGNED AND ARRANGED BY
CHARLES CAMPBELL, MEM. SAN. INST.

THE SOUTH KENSINGTON HYGIENE CABINET

Miniature sinks, lavatories and drains were used to train Sanitary Inspectors in good hygiene practices.

At first glance this cabinet could be mistaken for a child's toy. A rather elaborate construction kit, perhaps? But the real purpose of this set was deadly serious. Filled with an array of numbered models, some are clearly recognizable but others strange and unfamiliar. This Victorian cabinet illustrates an emerging understanding about public health and how domestic and work environments, and the behaviours of those who inhabit them, can have direct impacts on well-being. Rather than being toys, these are tools used to train a special group of workers known as Sanitary Inspectors.

Referred to as a South Kensington Lecture Cabinet, and probably one of a series, this object was constructed in the early 1890s. It was made to a design by Charles Campbell, of the Sanitary Institute (now The Royal Society for Public Health), which had been founded in the wake of the Public Health Act of 1875. This major piece of British legislation built on earlier regulations and followed years of pressure for improvements and reforms. Its regulations forced local councils to take greater responsibility in, for example, the provision of clean water, effective drainage and the removal of human waste. The law also called for the appointment of a Medical Officer of Health in every area. They in turn oversaw a team of Sanitary Inspectors, often women, who visited both homes and places of work in their local areas. There they would advise on good public health practice as well as report on breaches of it – particularly where health, or even lives, were at risk as a result.

In front of an intent audience, one might imagine a lecturer setting down and opening the cabinet, perhaps allowing the trainee inspectors to examine the contents as the concepts illustrated by the models were explained. Across three layers, it is packed with miniature water-supply and sewerage fittings, sanitary appliances such as toilets and wash basins, and a range of equipment relating to the ventilation of buildings. Both good and bad practices could be clearly demonstrated, giving the inspectors a taste of what they needed to look out for in the real, full-scale world around them. However, sanitation and ventilation were just two aspects of training that could also cover the fine detail of public health laws, diseases of both animals and people, building controls, and housing and factory regulations.

Opposite: The hygiene cabinet with some of its contents, 1890–95.

Left: One of the cabinet's larger models, showing a sink unit draining into a lavatory bowl, 1890–95. Not to be recommended.

AIDS AWARENESS LEAFLET

The British government launched a major information campaign about HIV and AIDS in 1987, including mass distribution of leaflets.

In 1987, as part of the biggest public health campaign ever undertaken in Britain, leaflets began dropping through the letterboxes of all its 23 million households. With measured language, these addressed some of the facts and dispelled some of the myths behind a growing health crisis. AIDS (acquired immune deficiency syndrome) covers a number of life-threatening conditions caused by HIV (human immunodeficiency virus). When it first made headlines in the early 1980s, there was no vaccine, no cure and little in the way of treatment to slow its progress. This led to considerable public unease and confusion. As news of the disease was emerging into societies where opinions and perceptions were largely shaped by popular media outlets, prejudice and stigmatization were also significant. This was particularly so as AIDS was initially presented as being associated with certain marginalized groups, most notably gay men and intravenous drug users.

In 1986, overseen by Health Secretary Norman Fowler, the British Government launched an HIV/AIDS-focused national health campaign. It was controversial from the start and faced significant opposition from within, including from then Prime Minister Margaret Thatcher. The country-wide leaflet drop was part of a broader strategy that incorporated newspaper and magazine adverts, billboard posters and a famous public information film, ominously narrated by the actor John Hurt.

One of the primary aims of the leaflet, like the wider campaign, was to cut through prejudice with the message that AIDS could affect anyone who was sexually active, regardless of their sexual orientation. It was also intended to dispel unfounded fears about shaking hands, sharing cutlery and other everyday interactions. However, some of those who criticized the campaign believed that it actually increased fears rather than improved understanding. It was also claimed that authorities had been slow to act in tackling the outbreak – which had initially spread rapidly amongst minority groups – leaving charities and pressure groups to lead the early public health initiatives.

Three decades on, in Britain the prognosis for those living with HIV/AIDS has been transformed. Treatments that are both effective and affordable have dramatically improved life expectancy, and pre-exposure prophylaxis medication, or PrEP, can significantly reduce the risk of contracting HIV. Unfortunately, for those affected or more at risk, the stigma and prejudice associated with HIV/AIDS continues across much of the world.

Left: The leaflet was part of a wider campaign, which included large billboard posters like this one in Halifax, 1986.

Opposite: Each leaflet was delivered in a sealed envelope marked "Government information about AIDS", 1987.

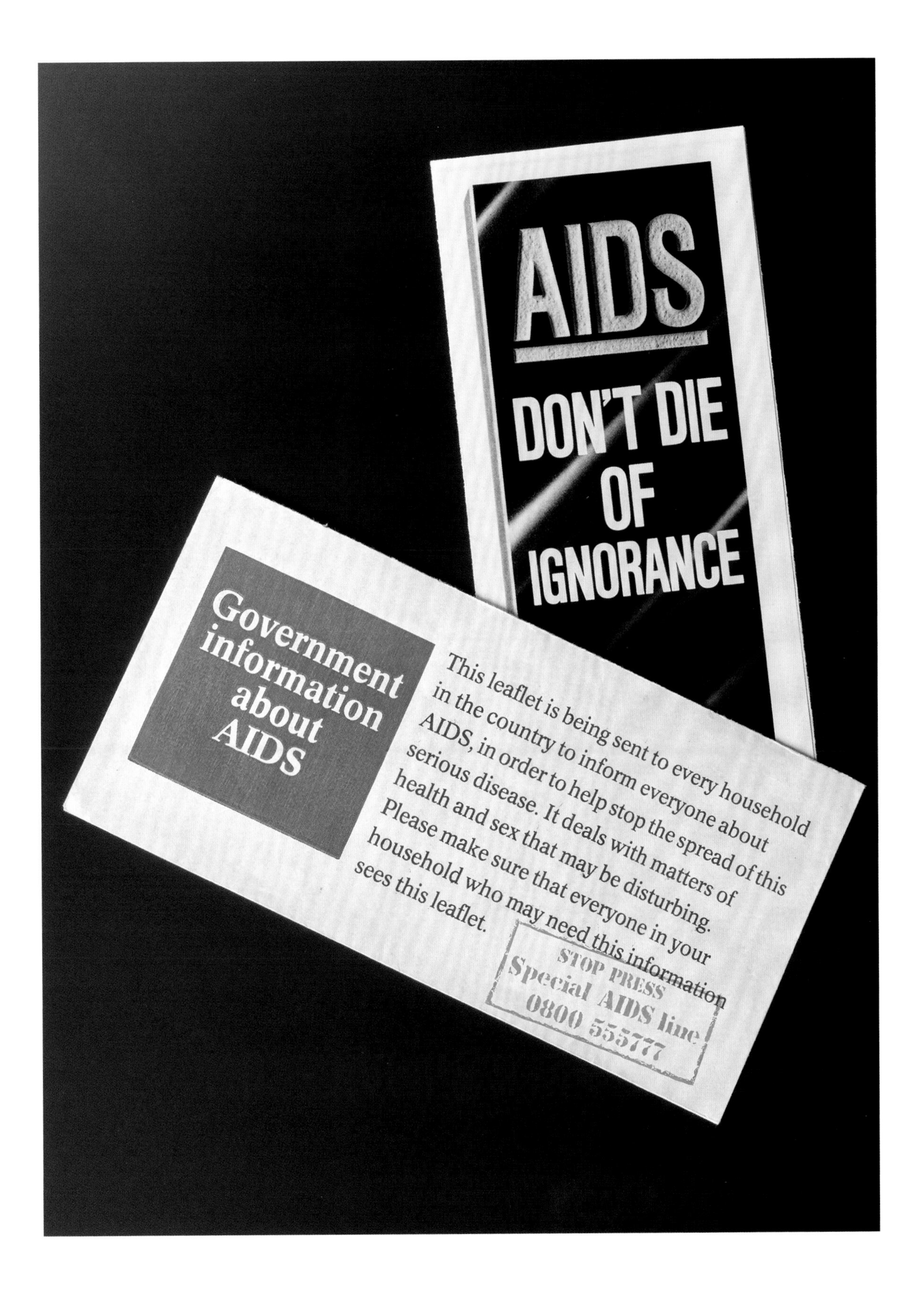
AIDS
DON'T DIE OF IGNORANCE
Government information about AIDS
This leaflet is being sent to every household in the country to inform everyone about AIDS, in order to help stop the spread of this serious disease. It deals with matters of health and sex that may be disturbing. Please make sure that everyone in your household who may need this information sees this leaflet.
STOP PRESS
Special AIDS line
0800 555777

THE BUG VAN

Portable fumigating vans killed insect infestations and infectious diseases from furniture during slum clearances.

Battered and weathered by years outdoors, the body of this van – which once sat upon a wheeled chassis – has undergone many hours of conservation work. Known as bug vans, this and others like it were once familiar sights on the poorer streets of towns and cities across Britain. They collected property from homes infested with bed bugs or other insect pests, or that had been the site of an outbreak of disease. Like the original horse-drawn versions in the late 1800s, these vehicles were commissioned, staffed and maintained by local councils.

The inner surface of this van is lined with zinc, which allowed for easy cleaning or disinfecting once it had finished its work transporting material from households struck by dangerous diseases such as smallpox. Depending on the circumstances, objects might either be incinerated or be taken to dedicated council sites for cleaning. Here they would be loaded into large steam disinfectors before being returned to their owners. Quite often the affected house would also be sealed up and fumigated as part of the cleaning process.

Having a bug van outside your home was undoubtedly perceived as a shameful event. It acted as a public announcement that a household had failed in an almost impossible task: to keep free of vermin or disease. Occupants could be seen as posing a health threat to their neighbours. However, the presence of the bug van could also sometimes have more positive associations:

"We were allocated a house ... it had a garden front, side and back. The council moved our furniture and belongings in the traditional way for areas like ours, namely in the green 'bug van' where everything was fumigated."

Norman Harding, Leeds resident, 1940s

The clearing of slum housing became a nationwide objective throughout much of the twentieth century. The people who lived in the slums were relocated into various social housing schemes, many of them maintained by local councils. Before moving, a family's possessions had to be fumigated to avoid spreading the infestation to their new home. In such circumstances the arrival of the bug van might have been seen as a symbol of change and perhaps represented the hope of a better life.

Left: In the 1930s in Bermondsey, a south London borough, this bug van was repurposed as a mobile cinema to screen public health films on some of the capital's poorest streets.

Opposite: This bug van was employed in the south London borough of Lambeth, probably in the years between the two World Wars.

NATIONAL ABORTION CAMPAIGN BADGES

People campaigning for a woman's right to access safe abortion created badges to share their message.

Badges have often been associated with political activism. They are inexpensive, mobile and easy ways to disseminate messages and raise awareness. They also allow the wearer to show solidarity for a particular cause, be this anti-war, anti-nuclear or in this case, pro-choice.

These protest badges were produced to support abortion rights for women. They formed part of campaigns to defend the 1967 Abortion Act – which made abortion legal on the proviso that two medical practitioners gave their approval and that a medical professional carried out the procedure – when further restrictions were proposed throughout the 1970s.

Abortion is an emotive topic, eliciting impassioned responses from pro-choice advocates and anti-abortionists alike. Campaigning is commonplace on both sides of the debate, resulting in a proliferation of ephemera; placards, banners, sloganed t-shirts and badges have populated these protests since the 1970s. These specific badges are linked to the activities of the National Abortion Campaign (NAC). It was established in 1975 to fight for abortion rights for women and to secure better access to the procedure in the face of opposition. The "March for Abortion Rights" badge dating from 28 October 1979 relates to a demonstration that took place in response to the 1979 John Corrie Amendment Bill, in which this former MP for Bute and Ayrshire proposed further restrictions to abortion. In this instance the NAC formed a subgroup, Campaign Against Corrie, and was eventually successful in getting the bill dropped.

Under current UK legislation, it is still not possible for a woman to authorize her own abortion without the consent of two medical practitioners. In 2003 the NAC merged with the Abortion Law Reform Association to form Abortion Rights, an organization that continues campaigning about this issue today.

In 2017 the Abortion Act turned 50. According to recent surveys, three-quarters of British people support a woman's right to choose. Yet there is still a long way to go to secure equal access to this procedure globally. Abortion remains restricted and even illegal in other parts of the world. Where there is no regulated provision of abortion services, the mental and physical health of women is at risk. Campaigners continue to fight for a woman's right to choose both nationally and internationally.

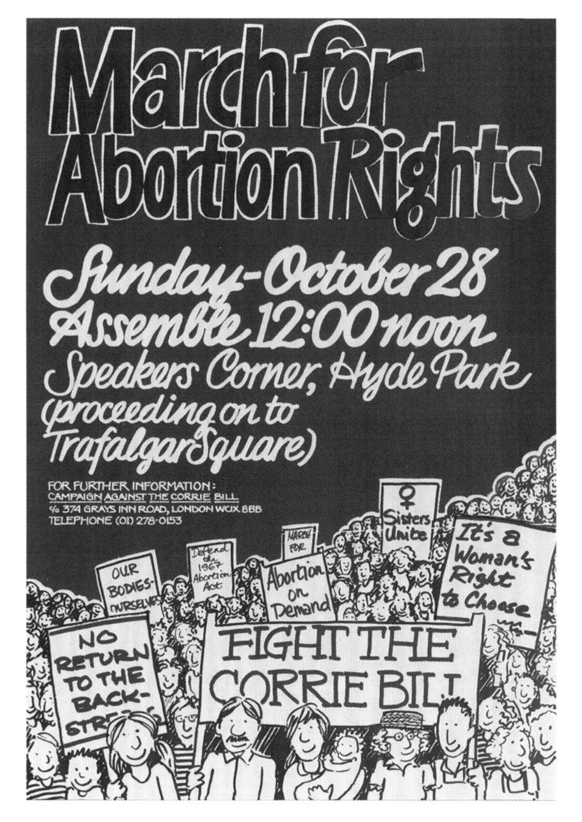

Above: Poster advertising a march for abortion rights, 1979.

Opposite: Five National Abortion Campaign badges, Britain, 1970–81.

A WOMAN'S RIGHT
abortion
TO CHOOSE
NATIONAL ABORTION CAMPAIGN
A WOMANS RIGHT TO CHOOSE
MARCH FOR ABORTION RIGHTS
OCT. 28
NO RETURN TO THE BACK-STREETS
FIGHT THE CORRIE BILL
A WOMAN'S RIGHT
abortion
TO CHOOSE
March for abortion rights
APRIL 3

Gratitude and Esteem
of his fellow townsmen
and unceasing attention to the
during the awful visitation of
MALIGNANT CHOLERA
AT
Plymouth
A.D. 1832.

A GIFT OF THANKS DURING CHOLERA

Surgeon Robert Fortescue received this silver snuff box in recognition of his work during the 1832 Plymouth cholera epidemic.

Bravery takes many forms, and so can the rewards given to the brave. Made by renowned silversmiths Nathaniel Mills & Sons of Birmingham, this snuff box was presented to a Plymouth surgeon in his early 60s, Robert Fortescue, in 1832. It marked the "gratitude and esteem of his fellow townsmen for his humane and unceasing attention to the poor during the awful visitation of malignant cholera".

Cholera was new to Britain and, after the first recorded case in Sunderland in October 1831, it swept across the country killing many thousands over the following months. It was the first of four waves of the disease that would strike the country during the Victorian period. The toll in Plymouth was particularly high. A port city with an expanding population, much of which was crowded into impoverished and filthy streets, the conditions were perfect for a disease we now know to be spread when bacteria are ingested, usually in food or water contaminated by human faeces. Nearly 800 people died there before the epidemic subsided.

Death and disease were commonplace in Victorian Britain, but cholera terrified the nation. Accompanied by severe vomiting and diarrhoea, the shocking physical transformation of the victims, who often turned a bluish grey, was matched by the speed with which they succumbed. "Well at breakfast and dead at suppertime" became a familiar characteristic of its visitation. And in 1832 there was no cure, effective treatment or understanding of how cholera was spread. Despite such dangers Robert Fortescue stayed the course as the epidemic played out, tending to the sick and dying in the poorest, most stricken parts of Plymouth.

Records suggest that Fortescue died in 1845, two days short of his 74th birthday. Four years later cholera returned to his home town. While the city's population had continued to grow, living conditions for many had deteriorated further. A report commissioned shortly before the second outbreak revealed, among other public health horrors, that in one courtyard 171 people were crammed into six houses, all sharing a single standpipe for their water supply. By the end of 1849 cholera had claimed another 800 Plymouth residents.

Opposite: The silver snuff box presented to Robert Fortescue in 1832.

Left: A young female victim of cholera, shown before and after contracting the disease, printed in Vienna, 1831.

SMOKING AWARENESS POSTER

Health campaigners used striking graphic art to inform the public about the dangers of smoking.

No research programme in the modern study of the environmental causes of disease is quite as significant as the correlation, established from the late 1940s onwards, between tobacco smoking and lung cancer. It is perhaps a surprise now – 70 years after Richard Doll and Austin Bradford Hill's first paper on the connection – that it took about 20 years for significant numbers of people, even doctors initially, to pay attention to the evidence and give up smoking. Making connections between popular "lifestyle habits" and serious disease was a new aspect of medicine at the time; such correlations were more usually seen in the hazards of working conditions in certain trades and industries. But health risk is now an enduring part of how we think about how to live our lives. This can be traced back to the slow revolution in public attitudes after the correlation was established, and as it became increasingly strong, implicating tobacco in many conditions other than cancer.

These posters are a tiny component of the enormous and complex cultural phenomenon that helped so many people in the West turn their back on smoking. The modern, public-health activity of health education and promotion was, effectively, a product of the First World War era, the same period that made cigarette smoking a mass phenomenon. Before this date Western tobacco smoking, which dates back to the seventeenth century, was normally a matter of smoking a clay pipe. With the foundation of the UK's Ministry of Health in 1919 the agenda of public-health doctors made a resolute change, building on nineteenth-century engineering solutions in public health, such as clean water supply and sewerage, to target people's personal habits and behaviour. Voluntary health associations ran the first campaigns, with the aim of reducing the incidence of tuberculosis, sexually transmitted diseases, even "uncleanliness". But it was only really in the 1950s that smoking and health education came together. In the case of the poster illustrated opposite, graphic art fought fire with fire by adopting some of the glamour of cigarette advertising. Other approaches – such as the Health Education Authority's memorable "No wonder smokers cough" poster (left) – came close to scare tactics to persuade people of smoking's danger.

Opposite: Public health poster warning against the dangers of smoking, 1990s.

Left: Poster probably produced by the Central Council for Health Education, London, England, *c.*1957–65.

Cancer

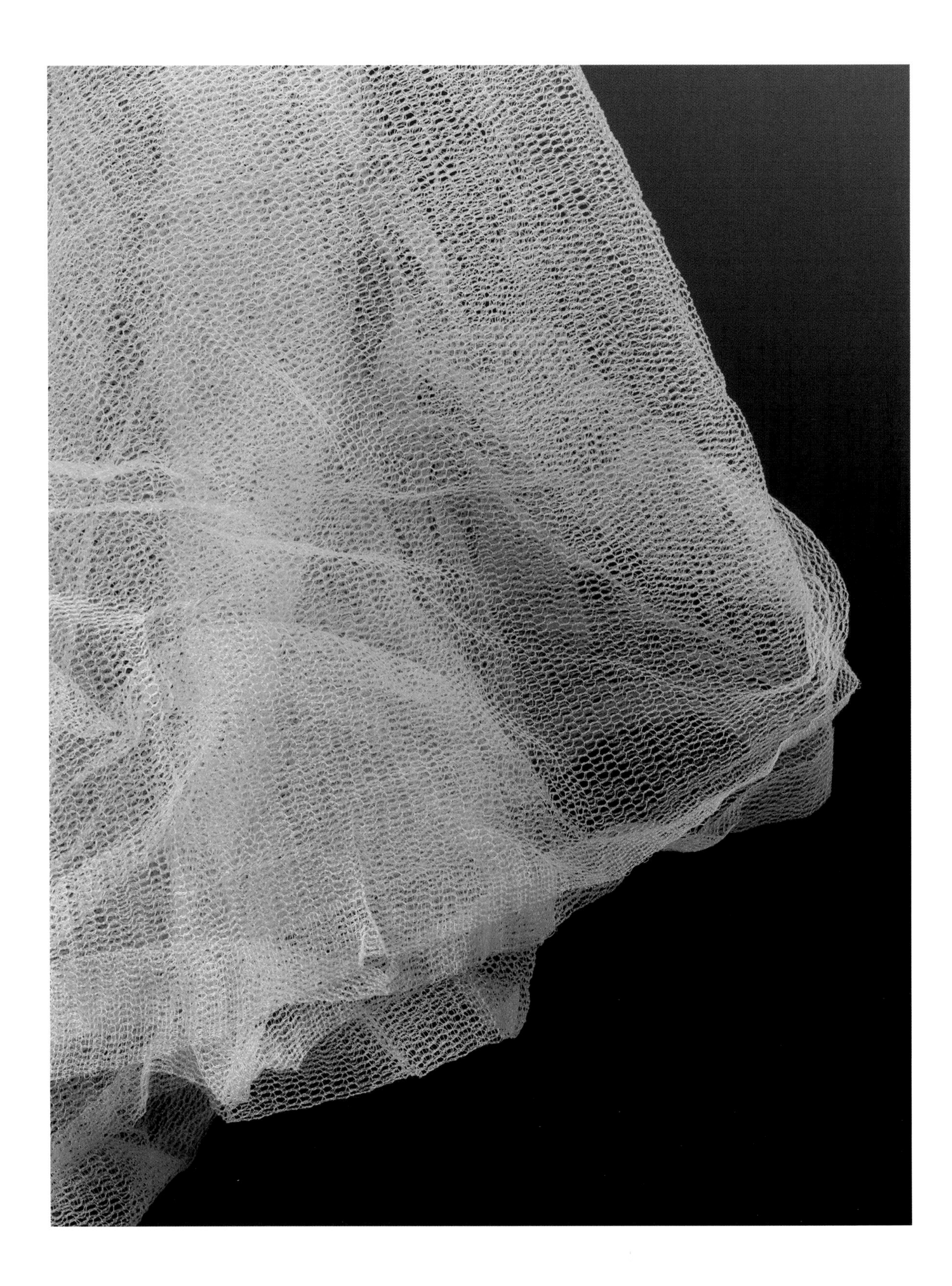

THE OLYSET MOSQUITO NET

A mosquito net, impregnated with insecticide and hung over a bed, provides the one of the most effective protections against malaria.

Mosquito nets offer protection against mosquitos, flies and other biting insects. They protect people by creating a physical barrier and are traditionally designed to hang over a bed at night.

The theory of infected mosquitos as a mode of transmission of malaria was proven by Ronald Ross in 1897 when he dissected the stomach tissue of a mosquito that had previously fed on a malarious patient and found the malaria parasite. This later led researchers to discover it was only female mosquitos that carried the parasite. This landmark discovery informed the basis for malaria prevention. Fine-mesh nets have long been used as a barrier against mosquitoes, but this example has an added layer of protection. The Olyset Net is coated with an insecticide called permethrin, providing both a physical and a chemical barrier against the malaria-carrying mosquito. The insecticide is said to repel and even kill mosquitoes. This, combined with the tough polyethylene fibres of the plastic weave net, makes it a durable product suitable for conditions throughout the world.

Sumitomo Chemical Co. of Japan developed the Olyset mosquito net in 1978 and the World Health Organization approved it in 2001, before going on to support its mass production as part of an international anti-malaria campaign in 2006. The net is classified as a long-lasting insecticidal net (LLIN) and its chemical coating is said to remain active for a minimum of five years, even after washing. The insecticide permethrin had previously been used for decades in public-health applications such as medically approved head lice shampoo. Permethrin is a powerful chemical but one considered safe for humans, posing minimal toxic risk to adults and children. With Olyset's controlled-release technology, the insecticide slowly works its way to the surface of the product throughout its lifespan, maintaining an effective surface concentration.

The combination of two long-established methods, nets and insecticides, has proven remarkably effective in the global fight against malaria. Since its inception the Olyset Net has saved an estimated 600,000 lives and averted approximately 100 million cases of malaria worldwide. Though this product uses hardwearing fibres and cutting-edge insecticide technology, it remains relatively cheap to buy. More than 300 million Olyset nets have been delivered to 80 countries worldwide, placing this product at the frontline of malaria prevention.

Opposite: Olyset mosquito net, Japan, 2005.

Above: Ronald Ross the year he discovered the link between malaria and mosquitoes, 1898.

CAECAL CONDOM

People used condoms made from animal gut to protect against venereal disease as well as a form of contraception.

Sheaths for the penis have existed since ancient times, created from materials such as leather, silk and oiled paper. What is less clear is whether they were initially used as a form of dress or have always been employed as protection against pregnancy and infection from sexually transmitted diseases. The first written mention of condoms is credited to Italian anatomist Gabriello Fallopio in Italy in 1564, although by the eighteenth century they were viewed as an English invention. The condoms mentioned by Fallopio were made of linen that had been soaked in salts, rather than the animal-gut variety illustrated here. According to legend animal-gut condoms were invented by a Dr Condom, physician to Charles II, in a vain attempt to slow the tide of illegitimate offspring of the English King. However, it seems that Dr Condom is a shadowy figure, and there is no evidence to suggest he is more than an urban myth. Even the origins of the word are unknown; research suggests a connection to the French town of the same name is unlikely.

This particular condom dates to the early twentieth century and was supplied by the pharmacy John Bell & Croyden of Wigmore Street, London. Jacob Bell, son of the original owner John, founded the Royal Pharmaceutical Society of Great Britain in 1842. The historic frontage of the family's shop was collected by Henry Wellcome after the pharmacy moved from its original location in Oxford Street in 1909, and is still held at the Science Museum today.

Animal gut condoms were made of either sheep or pig intestines from the section of gut that has a blind end, known as the caecum. The gut was soaked in water for some hours, turned inside out and immersed for several days in a weak alkali solution that was changed every 12 hours. It was then carefully scraped to remove any of the mucous membrane, before being exposed to burning brimstone, washed with soap and water, inflated for drying, then finally cut to the required size. Lastly the end would be hemmed and threaded with a red or pink ribbon that could be tied on for extra security. Some condoms were decorated; soldiers liked them emblazoned with their own regimental colours. Until the twentieth century condoms were used primarily as a protection against sexually transmitted disease rather than as a contraceptive – and therefore not to be used by respectable married couples.

Left: Two caecum condoms hang next to the door of the dying harlot, as two doctors argue about her treatment, from William Hogarth's *The Harlot's Progress*, 1732.

Opposite: Caecal condom, supplied by John Bell and Croyden, England, early twentieth century.

SARAH
OVER
READ
Homesick

VOLUNTEERS' DESK FROM THE COMMON COLD RESEARCH UNIT

Volunteers keen to help research stayed in the Common Cold Research Unit to catch snuffles and sneezes.

When the Common Cold Research Unit (CCRU) just outside Salisbury, UK, closed in 1989, the Science Museum was very keen to collect a representative sample of artefacts. The CCRU was a nearly unique experiment in clinical science which, by its very nature, became familiar to the public in a way uncommon to other aspects of medical research. The work of this unit of the Medical Research Council was to research the mechanisms by which colds were contracted and spread. The economic argument for its establishment related to the number of working days lost to colds, and the technique required the recruitment of healthy volunteers to risk being infected by the doctors at the unit. Christopher Andrewes, David Tyrrell and others created a series of experiments that investigated the propagation of infectious agents in sneezes and, afterwards, the identification of the very many viruses, of different families, responsible for the broad lay category "common cold". Volunteers would come to the unit – built into a prefabricated hospital donated by Harvard University during the Second World War – for the best part of two weeks, during which time they might, or might not, contract a cold.

To ensure a steady stream of volunteers, it was necessary to ensure that the unit featured in films, TV and radio programmes, and magazines and newspapers. A side effect of this was the establishment of an imagined public version of the unit that suggested a somewhat gimcrack enterprise, not entirely separate in the public mind from the activities of the Microbiological Research Establishment, which was located on the other side of Salisbury at Porton Down. The view the publicity put across was that, for many volunteers, attendance at the unit would simply turn out to be a good free holiday; if you were one of the ones who caught a cold, then that was a risk worth taking for the benefit of serving others by being a good citizen-guinea-pig. For all the jollity of stories about people spending their time making jam or enjoying country walks, however, the graffiti covering the inside of this desk from a volunteer's flat tells a different story – of the tedium of a stay at the CCRU. Here a museum object is an eloquent textual source for an untold history.

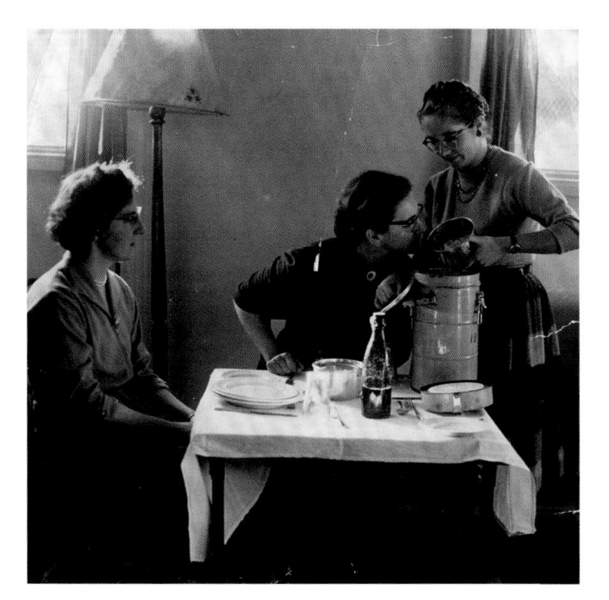

Opposite: Lino-topped wooden desk with fold-out section for use with typewriter, with extensive graffiti on the inside of drawers, from lounge of volunteers' flat at the Common Cold Research Unit, 1946–80.

Left: Volunteers living in isolation at the CCRU, 1959.

TEACUPS USED BY NATIONAL IMMUNIZATION DAY ORGANIZERS

Workers for National Immunization Day in Delhi needed refreshment to fuel their hard work in protecting children against polio.

Sometimes the simplest of domestic items can tell a tale of heroism and hard work. The cheerful, floral pattern of these teacups, and the even more welcome beverage they contained, sustained countless volunteer workers on the Polio Eradication Campaign in Madipur, south Delhi. A refreshing cup of tea was never better deserved.

National Immunization Day in India is an impressive feat of organization and diligence. Some 170 million children under the age of five are vaccinated against poliomyelitis (polio) twice each year by health workers based in temporary polio booths, set up in public spaces throughout India - often these booths are just a simple banner in a health centre, bus station or town square. Children are brought to the booths by their relatives to be given two drops of oral vaccine, then have their finger marked with an indelible ink mark. Following the mass vaccination at the booths, the detective work begins. In the four days that follow National Immunization Day, the vaccinators travel door to door, tracking down and immunizing the 70 per cent of children who had not been vaccinated at the booths. It takes five days to ensure that every single child across India is protected against the dreaded polio.

Previously known as infantile paralysis, polio is an infectious disease transmitted by faecal matter entering the mouth. Most cases show no or only mild symptoms, but for those who develop muscle weakness and paralysis, the effects can be devastating or even fatal. When an effective vaccine was discovered in 1955, it made headline news around the world.

The immunization campaign in India is an administrative triumph taking months of preparation. Doctors, health workers and volunteers store and maintain the vaccines, compile information to match the numbers of children with those vaccinated, and organize maps and any local information needed. A sweet and milky cup of tea gives a much-needed boost of energy to those determined that the wasted limbs and deaths of children contracting polio should become a distant memory.

Left: A local school promoting the polio vaccination campaign, Delhi, 2017.

Opposite: Teacups from Madipur Maternity Home, 2000s.

Russovia

6

ASSISTIVE TECHNOLOGY

Perhaps in no other area can the sheer range of human ingenuity be demonstrated more clearly than in the creation of technology designed to assist the limits of the human body. Assistive technology can be complex, having both cosmetic and medical elements as well as making life easier for its user. The humble pair of spectacles, for example, can be a fashion statement, add a corrective lens following eye surgery or simply help with reading. Prosthetics to replace missing limbs have been carefully created at home using the simplest of materials and crafted in prisoner-of-war camps using scrap metal; they can also showcase the best design and technology in the world, able to replicate the finest of human motor skills. Machines can replace the function of failing kidneys, enable people to move about more easily in many ways and even help with creating life itself.

THE PIANIST'S ARTIFICIAL ARM

Elizabeth Burton's artificial hand and forearm were specially adapted to allow her to pursue her career in music.

Few medical technologies are as intimate as an artificial limb. Fundamentally they are a physical manifestation of an absence, be it through war or accident, disease or genetics. But this material culture can also help reveal personal stories, as evidence of a desire to improve mobility, regain personal confidence or return to employment. Many of the historic prostheses within the Science Museum's extensive collections can still evoke the lives and experiences of their former owners. They can also help illuminate the medical realities of earlier times.

The one-time wearer of this very special artificial arm was Elizabeth Burton. Born in the 1860s, she became a teacher of music and singing and was based in the East Midlands town of Northampton. A family story tells of how, when preparing raw fish, she accidentally ran a fishbone deep beneath her right thumbnail. An infection set in and, as this occurred several decades before the development of antibiotics, the limb had to be amputated to save her life. Her hand and much of her forearm was removed.

Despite such a drastic loss, it is believed that this prosthesis allowed Elizabeth to continue her musical career. Made in 1903 by a local surgical-instrument maker, the arm is constructed from wood and leather, and has a large wing nut enabling the hinged, metal elbow joint to be locked into position – a useful feature when playing the piano. The hand can also be detached, suggesting that a more functional or perhaps more natural-looking prosthesis could be clicked into place when Elizabeth was away from the keyboard.

For a piano teacher, losing a right hand presents a special challenge as it traditionally plays the melody in music aimed at beginners. But as a mother of six, widowed in her 30s, the need to continue in her chosen profession may well have been critical, and it seems as though Elizabeth may have succeeded; records tantalizingly suggest she wore this arm when playing in concert at London's Royal Albert Hall in 1906.

Above: Elizabeth Burton, in a studio portrait with her youngest daughter Ada. Her right arm remains hidden.

Opposite: Artificial arm made for Elizabeth Burton in 1903.

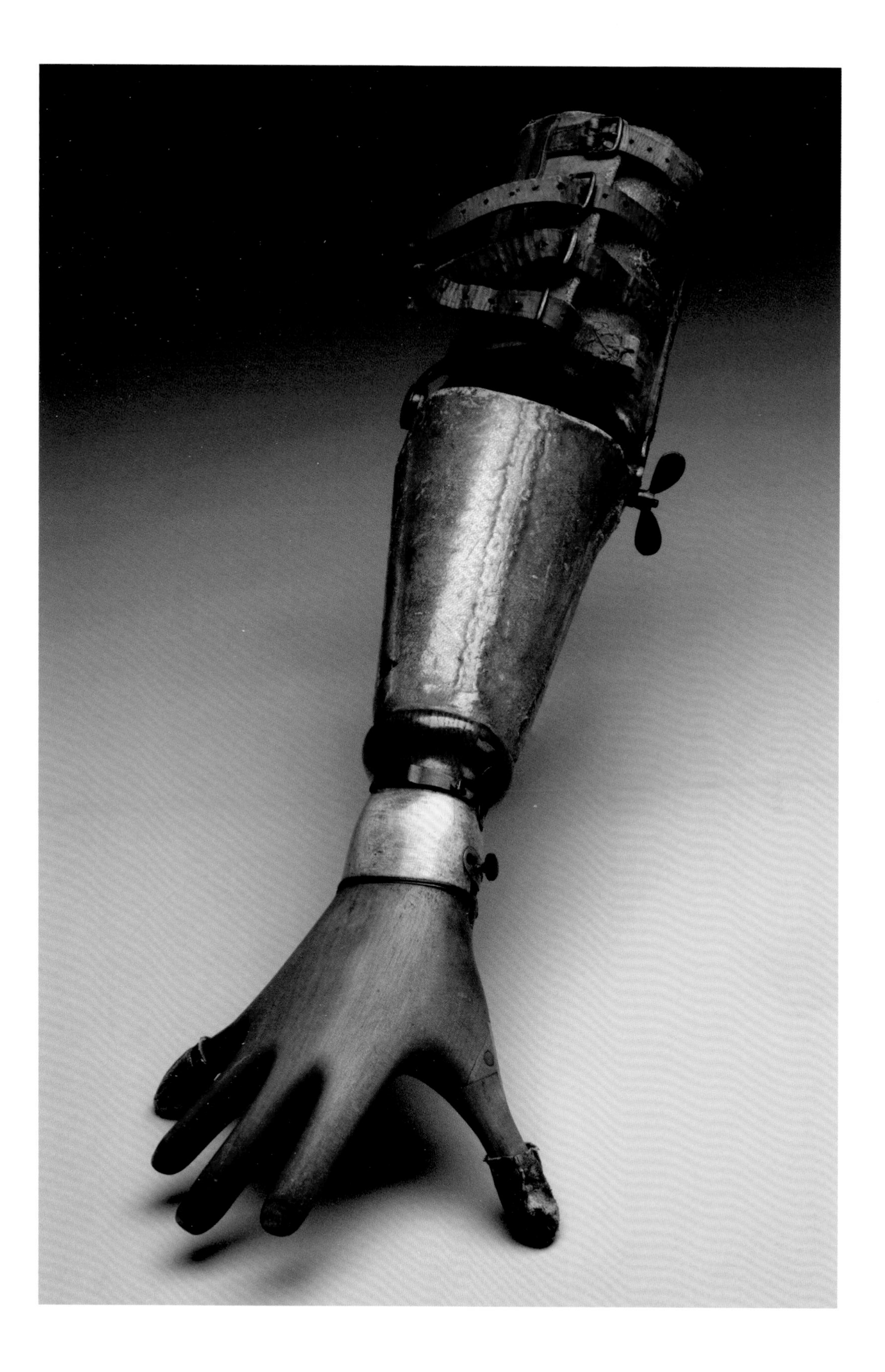

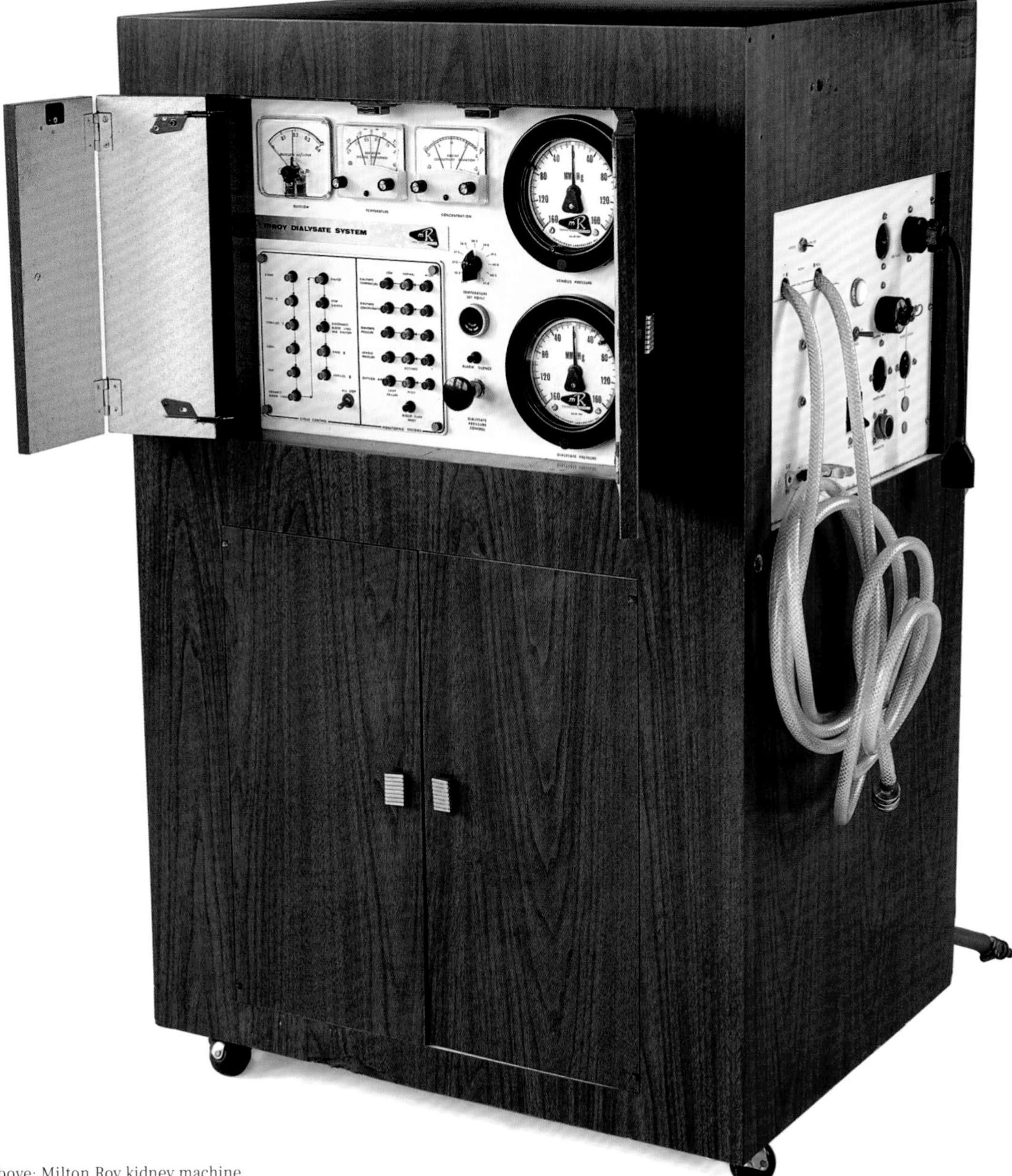

Above: Milton Roy kidney machine, used by Moreen Lewis, 1960s.

Opposite: Patient and nurse with dialysis machine, Royal Free Hospital Renal Unit, London, 1968.

MOREEN'S HOME-USE KIDNEY DIALYSIS MACHINE

Moreen Lewis was one of the first patients to have dialysis at home. She nicknamed the kidney machine "Dr Who".

Hidden inside a piece of household furniture, at first glance this medical equipment might appear mundane and unassuming. Throughout the 1960s and 1970s dialysis machines like this one were used at home by patients experiencing chronic kidney failure. Healthy kidneys function as the body's filters, removing harmful waste products from the bloodstream. If the kidneys fail, dialysis replicates this function by passing the patient's entire blood volume through an external machine, which filters out any toxins, before pumping blood back into the body.

Before home dialysis became possible, the procedure had to be performed by trained hospital staff in specialized renal units. There were only a limited number of beds available, meaning that new patients could not always receive this life-saving treatment. To resolve this problem, a US team led by Dr Albert Babb pioneered the first home-dialysis machine in 1961, which later became the Milton Roy haemodialysis machine. As it was operated by patients in their own homes, hospital bed-space could be freed up and treatment made more widely available. It also gave patients more independence from the hospital and helped to integrate dialysis into their everyday lives. It was even designed to look like a piece of furniture that would slip seamlessly into their homes.

The Milton Roy haemodialysis machine illustrated here once played a critical role in a very personal story of survival, keeping a woman called Moreen Lewis alive for nine years. In 1967 Moreen became one of the first British patients to use one of these home-dialysis machines, having been diagnosed with acute nephritis, or Bright's Disease, in 1959. Despite an initial belief that Moreen's death was imminent, her life was extended by nine years by this machine, which Moreen nicknamed "Dr Who".

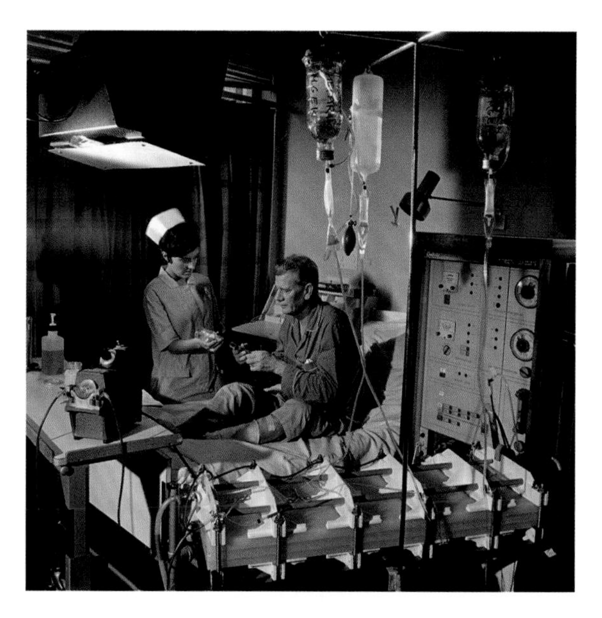

At a cost of £7,000 – twice the value of many houses at the time – independence came at a price, but this was fortunately raised by Moreen's family. Using the machine also brought an enormous amount of responsibility – not only for setting it up correctly but also for handling any potentially life-threatening leaks, breakdowns or haemorrhages that might occur throughout the 10-hour procedure. Fortunately, Moreen developed a great deal of self-confidence in using the machine, thanks to extensive training and practice, and her trust in the doctors and nurses at the National Kidney Centre, who were always just a phone call away.

INCUBATOR USED FOR THE WORLD'S FIRST IVF BABIES

Conceived outside the human body, the world's first "test-tube" baby started life in an ordinary glass incubator.

It looks like a mundane item of laboratory equipment, but this football-sized glass jar played a critical role in the creation of the world's first "test-tube" babies. In fact, no test tubes were involved; the first children conceived using *in vitro* ("in glass") fertilization, or IVF, began life in a Petri dish. The delicate embryos were then left to develop for several days, before being transferred into their mothers' wombs. So where did this object – an inexpensive desiccator jar – come in? Such vessels are routinely used to dry or protect their contents from water damage. This one, however, performed a more extraordinary function. An off-the-shelf model, it was repurposed by IVF pioneers Robert Edwards and Jean Purdy as an incubator for newly fertilized human eggs. Originally sealed at the top with a ground-glass stopper, it helped maintain the ideal environment for growth by replicating conditions inside the body.

Edwards, a British physiologist who had dedicated his career to the study of human reproduction, was the first person to fertilize a human egg outside the body, in 1968. Edwards joined forces with surgeon Patrick Steptoe, known for his pioneering use of laparoscopic or "keyhole" surgery to diagnose and treat gynaecological disorders – a technique he later perfected for collecting eggs from the ovaries. With their funding limited, Edwards and Steptoe relied on cheap or second-hand equipment and voluntary assistance from nurses to care for their patients. Purdy was one such nurse, whose laboratory expertise made IVF treatment a reality. A skilled technician, Purdy monitored the developing embryos and was the first to witness their cells dividing. The trio embarked on a decade of clinical trials involving no fewer than 282 women. Their work was met with widespread hostility from the scientific community, which argued that infertility was hardly a priority in an increasingly overpopulated world.

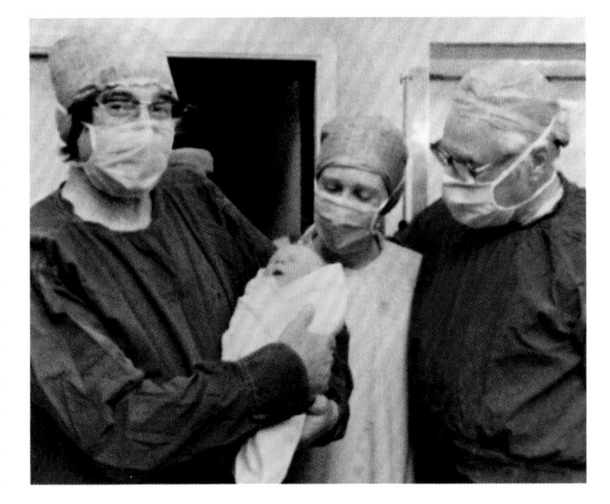

Above: Robert Edwards holding the world's first test-tube baby, Louise Joy Brown, with Jean Purdy in the centre and Patrick Steptoe on the right, 25 July 1978.

Opposite: Incubator jar used by Robert Edwards and Patrick Steptoe in the creation of the world's first "test-tube" babies, 1977.

Everything changed when, on 25 July 1978, the world's first IVF baby, Louise Brown, was born at Oldham and District General Hospital in Greater Manchester. Her arrival created a media sensation, overturning scientific and popular opinion and offering a beacon of hope to people otherwise unable to conceive. Since Louise's birth more than six million "test-tube" babies have been born. The team's achievements were finally recognized in 2010, when Edwards – the only remaining living member – was awarded the Nobel Prize in Physiology or Medicine for the development of IVF.

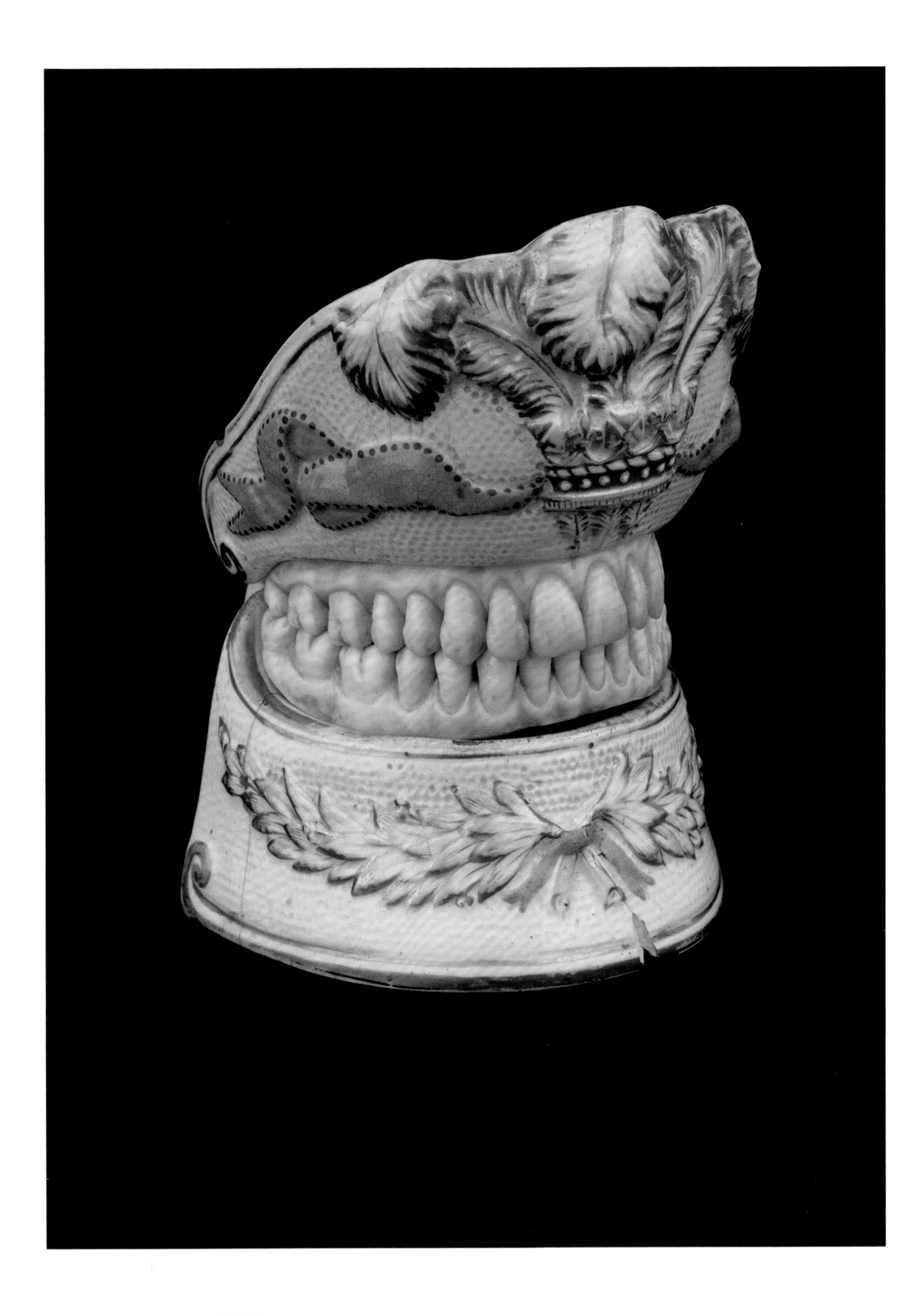

HIPPOPOTAMUS IVORY DENTURES

It needed an expert to carve dentures out of hippopotamus tusks, and a rich person to be able to afford them.

Made from hippopotamus ivory, these dentures are displayed on a grand ornamental ceramic stand decorated with three white feathers emerging from a coronet, the motif of the Prince of Wales. Might they have been made for or even used by the royal mouth of the Prince, later George IV?

The maker has carefully carved the ivory to the shape of the wearer's mouth to ensure that that the dentures stayed in place through suction. At this time artificial teeth were also made from walrus and elephant ivory, but hippopotamus was preferred due to its durability. Ivory deteriorated in the mouth over time, turned black and smelt very unpleasant. For this reason human teeth, taken from the dead or the poor, were sometimes attached to an ivory base instead. The time it took to make dentures, as well as the expensive material used, meant only royalty and the upper classes owned such sets.

The Prince of Wales was an ostentatious regent, who enjoyed the finer things in life; he was an arbiter of fashion, an avid collector of the decorative arts, and a voracious eater. Like other affluent Georgians, the Prince of Wales ate the sugary foods that became increasingly popular in the eighteenth century. The resulting tooth decay and tooth loss, in addition to growing societal concerns with fashionable appearances, resulted in a demand for false teeth. Bartholomew Ruspini was the Prince of Wales's surgeon-dentist around the turn of the century and may well have made these dentures.

Bone and gold dental prosthetics were produced by ancient civilizations as far back as the Etruscans in 700 BCE but Frenchman Pierre Fauchard was the first to write about making artificial teeth in the modern era, in his *Treatise on Teeth* in 1728. Fauchard coined the name of a new professional called the *dentiste*, who had scientific credentials and offered restorative dental care.

Today visits to the dentist are the norm and it is hard to imagine wearing such an uncomfortable object following tooth loss. Porcelain teeth and vulcanized rubber bases replaced ivory and human teeth sets in the latter part of the nineteenth century, and since the 1940s dentures have been made from acrylic. Good teeth have progressively become a signifier of health and beauty, and in their absence dental prosthetics continue to assist individuals in realizing this widely accepted bodily ideal.

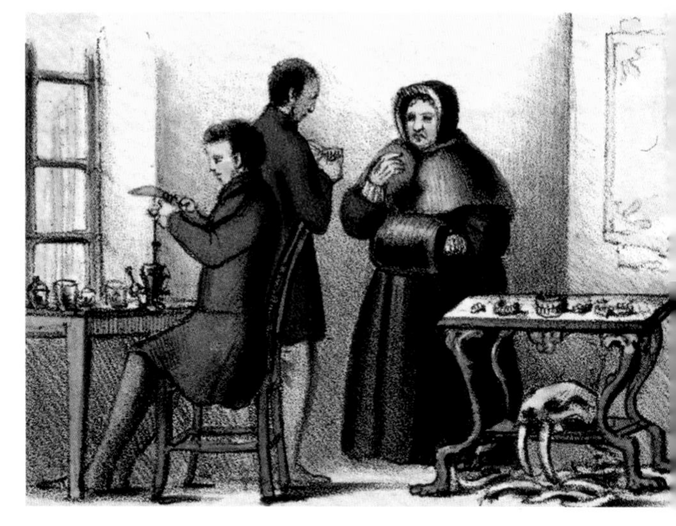

Opposite: Hippopotamus ivory dentures on a ceramic stand decorated with the Prince of Wales's feather badge, early nineteenth century.

Above: *Ivory for Dentists' use*, an illustration showing dentures being made from walrus tusks, by Benjamin Waterhouse Hawkins *c.*1845.

ED FREEMAN'S ADAPTED CAR

Edward Freeman had his orange Mini modified for his use and included a mini-bar for his friends at the same time.

Passing their driving test is a rite of passage for many teenagers – the freedom to go wherever they want to, at their own pace and in their own time. In 1978, for 19-year-old Ed Freeman, this bright orange Mini Clubman was exactly that. However, it wasn't his first choice of car – he preferred the more iconic Mini Cooper but that would never have fitted his wheelchair. Originally, the car fitters wanted Ed to have a high roof to accommodate his wheelchair to which he responded: "There's no way I'm driving an ice cream van!" A compromise was soon reached, and the floor was lowered instead, with a ramp for Ed to use. Like all teenagers at the time, he set about customizing his vehicle – with furry dice adorning the rear-view mirror and a mini bar installed for his passengers.

Ed Freeman is one of the hundreds of people in the United Kingdom who were affected by the drug thalidomide. Ed has shortened arms and legs, known medically as four-limbed phocomelia. His mother took two doses of thalidomide, prescribed by her local GP, while pregnant with Ed in 1959, to help treat her pneumonia. When Ed was born, the connection between thalidomide and its impact had not yet been made. The same GP asked Mrs Freeman if she had taken thalidomide – and she still had the remaining doses in her medicine cabinet to prove it.

Thalidomide was marketed by its makers as a completely safe sedative for all, with no fatal dose. It was sold and prescribed under 40 different names in 49 countries, for a range of conditions including morning sickness. Thalidomide was officially withdrawn from sale in the United Kingdom in 1962, after three-and-a-half years on the market. Stronger regulation and reporting systems were put into place across Europe, but the drug changed our relationship with medicines, and with medical professionals, forever. Ed and his family were among many who joined campaigns for compensation from thalidomide's makers.

Above: Ed Freeman as a teenager, pictured with his father, *c.*1978.

Opposite: A modified car driven by Ed Freeman.

Ed drove this car until 1988, when he switched to a bigger vehicle to cope with the demands of family life. This vehicle languished in Ed's garage until he donated it and other items, including the prosthetic legs he used as a child, to the Science Museum. He said: "Well, I don't want me or any of us just to be history, because then you stand a chance of repeating it."

YFL 693T

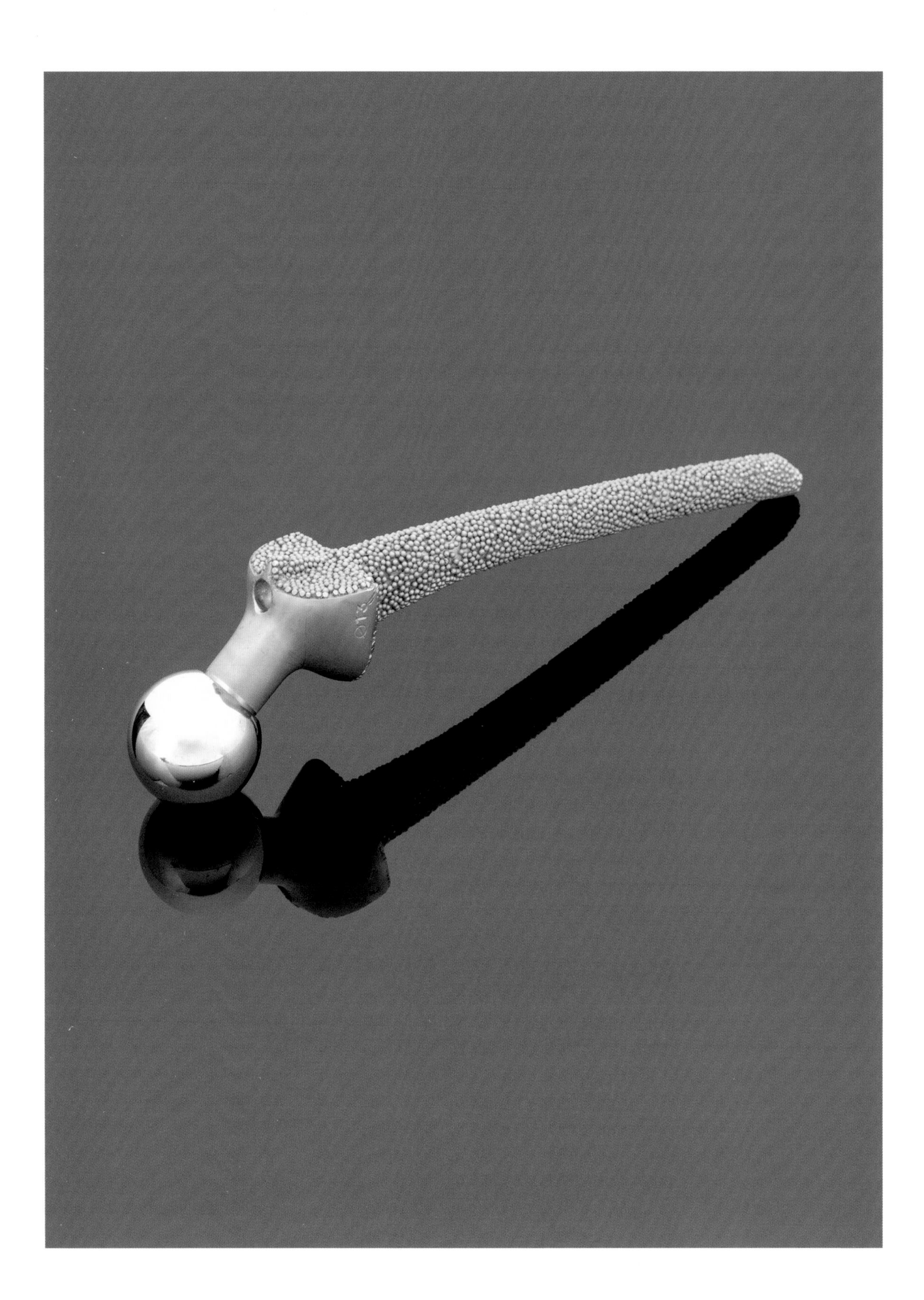

HIP REPLACEMENT JOINT

John Charnley was determined to create an artificial hip that moved silently with no annoying squeaks for patients with a hip replacement.

Imagine if your hip squeaked every time you moved your leg. This was the annoying reality for some patients who underwent hip replacement surgeries before the creation of the Charnley-type hip prosthetic in 1962. This new model was a result of the dogged determination of John Charnley, an orthopaedic surgeon, to improve the lives of hip replacement patients. This object was made of two parts; one was placed into the thigh bone of the patient, and the other cemented into their pelvis. Together they worked in the same way as the natural hip joint.

The impact of this invention cannot be overstated and has earned Charnley the title the "father of the hip replacement", but others had tried before him. Hip replacements were attempted in the nineteenth century, many of them in Westminster Hospital in London. Here the first ever excision arthroplasty (removal of the head and neck of the femur bone) was reportedly performed by Anthony White in 1821. But it was not until the 1890s that Themistocles Gluck replaced a patient's hip with ivory and fixed the prosthetic in place with nickel screws. This was the first known complete hip arthroplasty.

Through the early twentieth century the operation was carried out by a variety of practitioners, using acrylic, rubber or the alloy Vitallium. Though these joints improved mobility for patients in the short term, corrosion and degradation meant that many were immobile and in significant pain a few years later. After an encounter with a patient with a particularly squeaky hip, Charnley determined to design a lower-friction model. He tried a variety of different materials and methods to reduce the resistance inside the patient's hip, eventually putting polyethylene inside the socket. This allowed the joint to move naturally, resulting in less friction and less deterioration, and improving mobility for longer.

Opposite: Charnley-type hip replacement, with textured stem to aid bone grafting, 1990s.

Above: John Charnley, the "father of the hip replacement".

Charnley may not have been the first surgeon to replace a hip, but he was instrumental in creating a well-refined speciality. He was the first to introduce bone cement to seal the implant inside the patient, resulting in far greater stability, comfort and durability. His inventions improved the quality of life of thousands of hip replacement patients, so much so that Queen Elizabeth II knighted him in 1977 in recognition of his contributions to medicine and science.

PORTABLE INCUBATOR FOR PREMATURE BABIES

When most births happened at home, premature babies needed portable incubators to keep them warm on their journey into hospital.

In Britain today it is generally a mother's choice whether to opt for a home birth, birthing centre or hospital delivery. However, until the second half of the twentieth century, most babies were born at home. Portable incubators, like this 1960s example, were frequently needed for journeys to hospital by ambulance. This incubator was powered, via a transformer, by the transporting vehicle's batteries. It features controls to adjust temperature, and an emergency mask to attach to an oxygen cylinder, which was stored underneath the incubator; these were to prevent hypoxia (a lack of oxygen) and hypothermia (caused by low core temperature, which premature babies struggle to regulate). These incubators were also used to move babies to special care units from outlying hospitals. Given how expensive these machines could be, they were rented out by the company to local areas, and became a lifeline for communities.

During the 1950s, specialized baby care units were established by paediatricians, providing maximum support for premature or sick babies. The units isolated them from infection and allowed specialized care. Incubators such as this provided a controlled environment into which a premature baby was placed to monitor its progress and increase its chance of survival. Oxygen and carbon dioxide levels could be measured and adjusted, food administered and the temperature controlled. This incubator was made by the Oxygenaire Company, whose staff regularly attended neonatal meetings to ensure that their products were developed in tandem with paediatricians' needs. The company was a specialist in making oxygen therapy equipment and applied improvements to their portable and hospital-use designs. They also sponsored lectures at The Neonatal Society, formed in 1959 from the British Paediatric Association, to reflect the growing specialism of neonatal care.

By the 1980s, 90 per cent of births happened in hospital, so this type of portable incubator passed out of medical use. When it was collected from St Anne's Hospital in Bristol, curators believed it to be the sole example of this type of incubator left. Neonatal care continues to push the boundaries of the possibilities of medicine, raising questions about survival and the very definition of being premature.

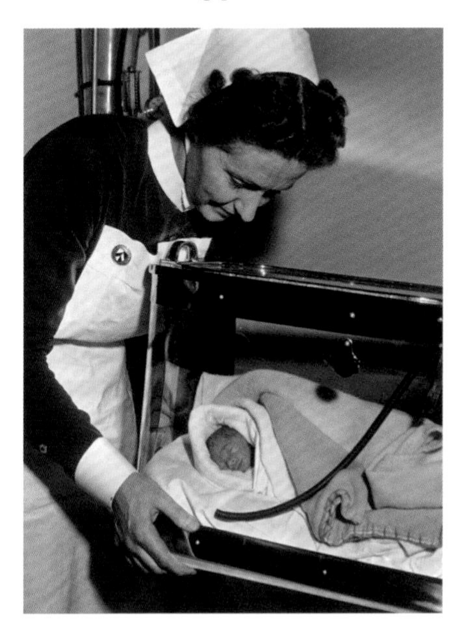

Above: Three-week-old premature baby being fed, 1958.

Opposite: Oxygenaire Series 3 portable incubator with separate transformer container, *c.*1960.

SAFETY ALARM IS
SET TO MAKE CONTACT
AT 98 DEG.F
Oxygenaire
OVERHEATING
LIGHT
UNIT
LIGHT
TEMPERATURE CONTROL
HIGH
LOW
INTERIOR
OFF
ON
LIGHT
UNIT
OFF
ON
SWITCH
FUSE
TEST
BUTTON
OUTPUT
Oxygenaire
TRANSFORMER
220/240 VOLTS
INPUT

DORRANCE
RIGHT
Model 5X

THE DORRANCE HOOK

After David Dorrance lost his hand in an accident, he designed the best artificial replacement he could by adapting earlier versions.

When it comes to technological solutions, simplicity is very often best. A perfect illustration of this maxim is provided by an invention patented by American David Dorrance in 1912. Having lost his hand in a sawmill accident in 1909, Dorrance set about improving the standard hooks that were available to amputees like himself. The result was the Dorrance split hook. More than a century on, refined variations on his original invention are still being produced, and it remains the most used prosthetic-arm terminal device in history. Peter Pan's fictional foe Captain Hook may have had his villainy enhanced by the threatening prosthesis that gave him his name, but hooks, albeit more practical ones, remain the choice of many limb-wearers despite all the other advances in artificial limb technology.

Dorrance's hook was all about function. It made no attempt to mimic the appearance of a natural hand and could not easily be disguised with a glove like many of the more cosmetic, functionless artificial hands of the early twentieth century. By splitting the hook, and with the addition of a thick rubber band, Dorrance's device allowed wearers to pick up, manipulate and even squeeze objects – be they everyday domestic items or the specialized equipment and machinery of work. Wearers opened the hook, in a scissor-like way, by means of body movements, which tightened a cable connecting one side of the hook to a strap looped around the user's other shoulder. When the wearer relaxed, the powerful rubber band then closed the hook again.

The date of the original patent gives a clue as to its rapid uptake and rise in popularity – within two years the First World War began. Amputees in their thousands began returning home to be fitted with artificial limbs and begin their rehabilitation. Many of these were young men, and most were desperate to return to some form of paid employment. Such was its versatility that the Dorrance hook became a favourite among veterans of this and later wars, be they labourers or office workers, as it gave them the opportunity for greater independence and the chance to take their place back in the workforce.

Opposite: A version of the Dorrance split hook, made in 1998.

Left: Double amputee and Oscar-winning actor Harold Russell lights his wife's cigarette with the aid of a Dorrance hook, New York, 1946.

THE YELLOW PERIL

Dan Everard created the "Yellow Peril" powered wheelchair for his daughter Ruth when she was aged 22 months.

What do you imagine when you hear the words Yellow Peril? Danger? A brightly coloured menace? Perhaps this is an apt description when you consider that this powered wheelchair, affectionately known as Yellow Peril, was driven by 20-month-old Ruth Everard in the 1980s. Using a joystick that gave Ruth control over its direction, speed and height, this chair helped her to explore the world around her.

At the age of 14 months, Ruth had been diagnosed with spinal muscular atrophy (SMA), which causes muscle weakness, meaning that she could not sit or stand without support. Despite this Ruth's parents were determined that she have the same level of independence as other children her age, and so began hunting for a tool that would improve her mobility. Frustration ensued. As put by Ruth:

> "My dad went looking for the tool that he knew I needed, and nobody had made it. He couldn't get it anywhere. The argument went:
> 'I need a wheelchair for my 18-month-old daughter.'
> 'Well, you can't put an 18-month-old in a wheelchair.'
> 'Well, why not?'
> 'Because nobody's ever done it.'
> 'Well, why has nobody ever done it?'
> 'Because you can't.'
> So he designed the thing that he couldn't buy."

This is the tool that Ruth's father Dan created – a powered wheelchair that Ruth started driving just a few months later. With more control over where she could go and what she could see and do, Ruth had the independence to stop and ask questions when she encountered things that she had not seen before, learning life lessons that otherwise could have literally passed her by in a regular pushchair.

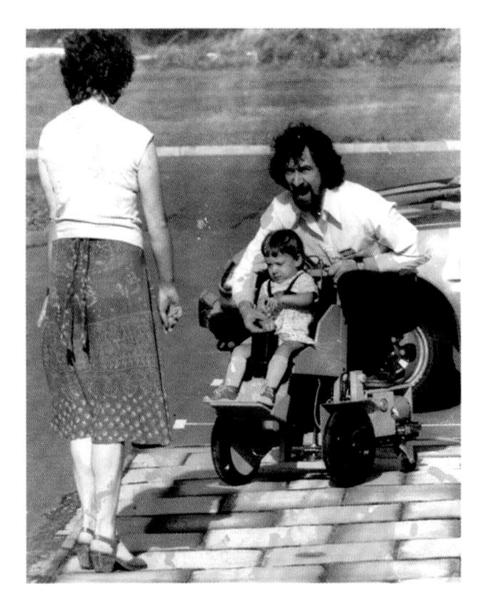

Opposite: "Yellow Peril" powered wheelchair designed and made by Dan Everard for his daughter Ruth aged 22 months, Cambridge, UK, 1981.

Above: Ruth using the "Yellow Peril", Cambridge, UK, 1981.

Why is this important? Childhood is a period where you actively learn about the world. Curiosity knows no bounds, and you constantly test your limits, find strange objects and learn what you can and cannot do. Sometimes quite literally getting a taste for life (to the horror of whoever is supposed to be watching you), you gradually begin to understand the rules that will keep you safe – and become independent enough to break them.

Since the 1980s the importance of power mobility aids for young children with reduced mobility has become widely recognized, and similar chairs have been developed by Dan and Ruth. The innovative Yellow Peril seems to mark the beginning of this journey to making equal participation, exploration and potential mischief-making more accessible.

RUTH
Russavia
DE HAVILLAND
89A

NATIONAL HEALTH SERVICE SPECTACLES

Spectacles are designed with an eye to design as well as their effectiveness as an optical aid.

National Health Service spectacles offer a fascinating insight into access to basic medical provision through the NHS, and the importance of appearance in the production of assistive devices.

Today spectacles are commonplace to improve our vision. However, for much of their history the cost of the frames and lenses, or of an eye examination, would have prevented most from experiencing their benefits. The introduction of spectacles to the NHS on its founding in 1948 was the first large-scale attempt to make them accessible to all. The number of wearers who took up these free versions made them a symbol for the newly established NHS and demonstrated its ability to provide access to basic medical care.

The demand for spectacles makes them an unusual example in comparison to other assistive devices. The number of spectacles required is far greater than the number of prosthetic limbs required. Moreover, the need to provide considerable quantities through a welfare service influenced the designs that were made available. Cost effectiveness and utility became important, and this was at the expense of the attractiveness and elegance of the frame. This meant that the spectacles were durable, but they were basic and designed on purely functional terms; fashion was of little importance.

NHS spectacles were produced in both metal and plastic, like the example shown here. However, they were distinct in style and could differ markedly from other frames developed in the wider commercial market that placed greater emphasis on appearance and user choice. This distinct style was often heavily stigmatized, highlighting social perceptions towards both assistive devices and basic medical provision at the time. How important would the appearance of a spectacle frame be to you? Is the design of the frame of more or less concern than the correction of vision itself? NHS spectacles are a useful example for exploring the design and social perceptions of assistive devices. However, their story also highlights the need for accessibility to vision correction and the difficulties of balancing this with cost in the modern world.

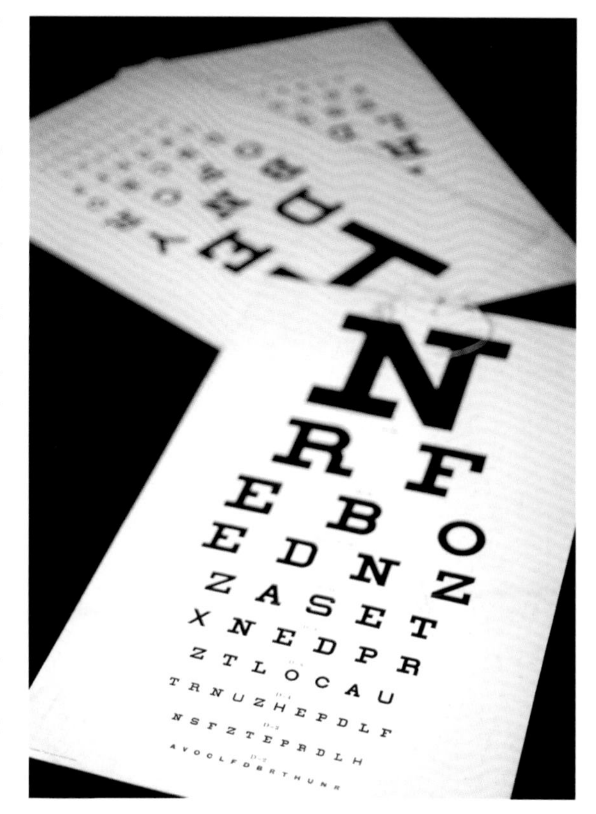

Opposite: Pair of plastic National Health Service spectacles, dispensed by Crispe the Opticians, Ruislip, *c.*1950.

Above: Eye testing charts, 1930–70.

14509.
LA LVNE
IANVARIVS
FEBRVARIVS
MARTIVS
APRILIS
MAIVS
IVNIVS
IVLIVS
AVG
SEPTEMBER
OCTOBER
NOVEMBER
DECEMBER

7

BELIEF

Sometimes science alone is not enough to heal the human body. The enormous complexity of the relationship between the mind and the body is barely comprehensible, but a patient's trust in their treatment, their faith and the support of their loved ones can have direct positive outcomes on their health. The power of faith and trust in promoting good health has an ancient history, with healing traditions featuring in many global religions. Performing a pilgrimage, drinking or bathing in holy water, or asking a deity to intervene in a cure are as commonplace today as they have been for centuries. Folklore can also feature remedies such as good luck charms and amulets as well as rituals, and these form an individual and creative record of the wish for good health. Examining the belief and faith that relate to healing provides us with a clue to understanding the human condition.

BRONCHITIS NECKLACE

Anxious parents believed that if their child wore a special beaded amulet, the child would be protected against bronchitis.

If you were worried that your child might have bronchitis in early twentieth-century London, a popular remedy was a necklace of glass beads. This beaded bronchitis necklace is part of a larger history of healing. It shows us that objects used to cure and protect have taken many different forms and have been invested with power from a variety of different sources. Even though their means of operation have not always been understood, and belief in their efficacy has changed over time, within their own contexts amulets have played a significant part in medicine since time immemorial.

This necklace is one of several acquired by English folklorist Edward Lovett. He had a profound interest in charms and amulets (terms which today are often used interchangeably), collecting a large variety from the city's markets throughout his life. Lovett visited 26 districts asking for a remedy for a child suffering from bronchitis and created a map as a visual record of all the places he was offered a beaded necklace like this one. While the objects' colours varied – often blue and green, but sometimes multi-coloured – their function remained the same: acting as a cure for or protection against bronchitis. They were worn to be invisible, often hung at the neck or beneath the collar of one's clothes. According to Lovett, necklaces such as these were often kept on from childhood right until death, and their efficacy depended on constant contact with the body, a feature common to amulets.

Amulets are objects that have been invested with power to cure or protect against illness and misfortune. Their form and function have varied hugely across time and space, and ranged from mundane, everyday and personal things, to expensive, mass-produced or culturally recognized items like the four-leaf clover or the Evil Eye. An amulet's potency often derives from the inherent properties of its materials, but it can also be imbued with power through physical alteration such the addition of inscription, images or symbols.

Historically across Europe children have worn necklaces as a form of healing; commonly for teething, and often made from materials like coral or amber. Did beaded bronchitis necklaces form part of this long-held medical tradition? Whatever the origins of its function or power, the medicinal value placed on items like this green necklace by London's inhabitants is clear; as illustrated by the map, they were widely used and known in the capital as a powerful remedy for bronchitis.

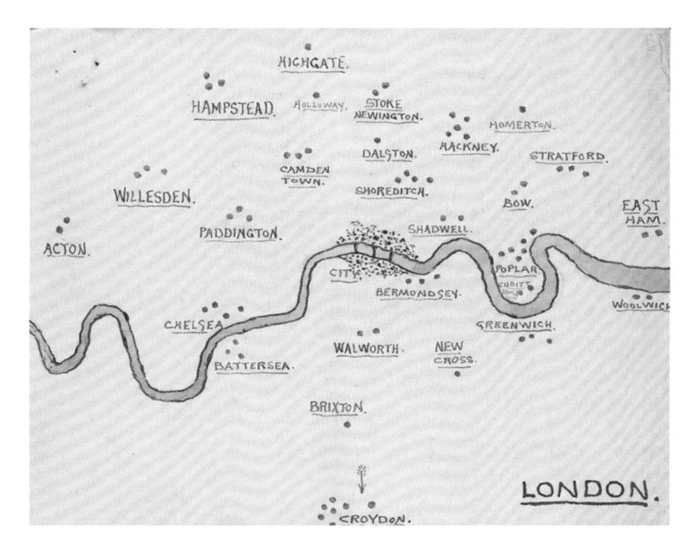

Left: Map created by Edward Lovett showing districts of London in which he found beaded bronchitis necklaces for sale, 1920s.

Opposite: London bronchitis amulet, early twentieth century.

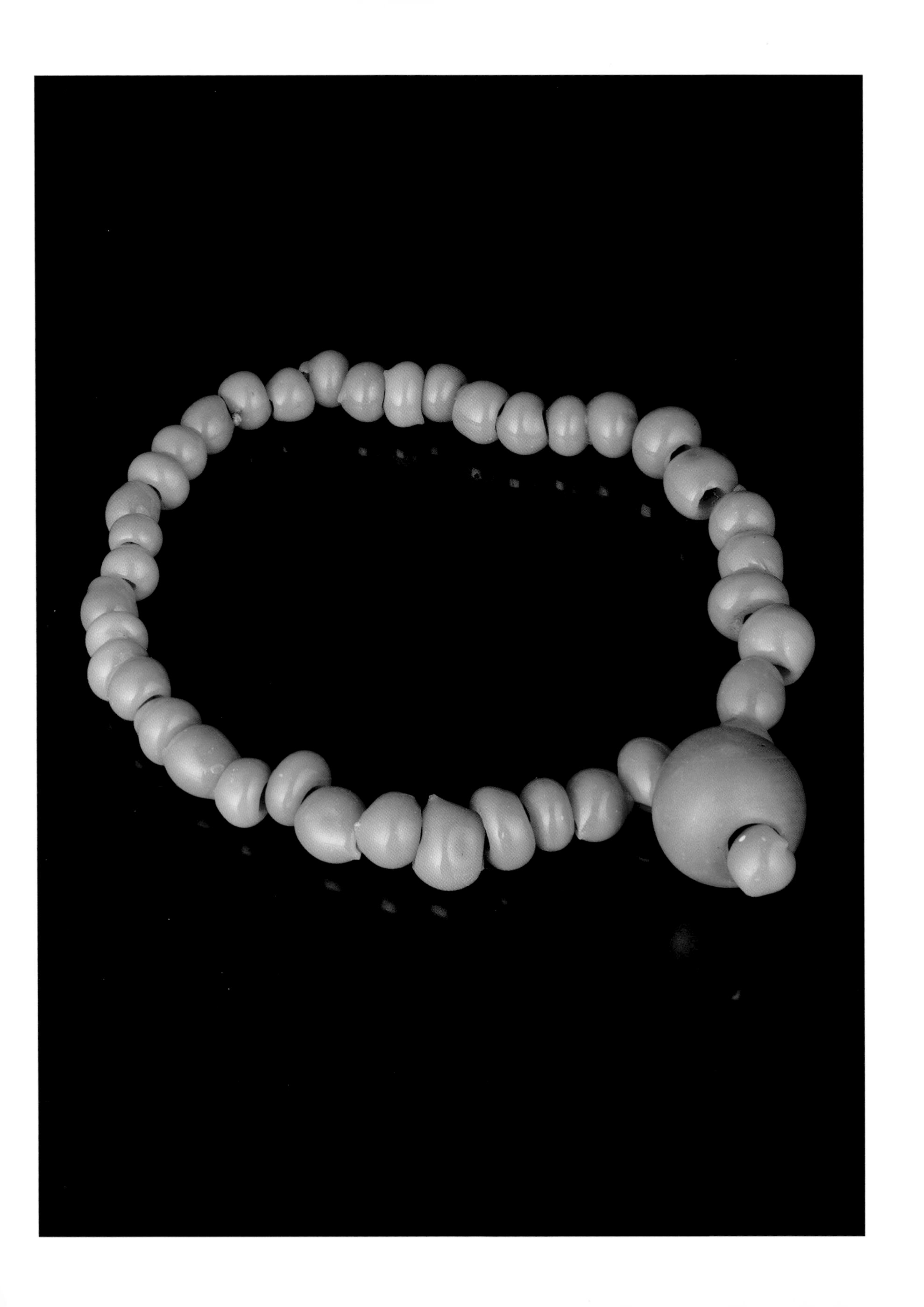

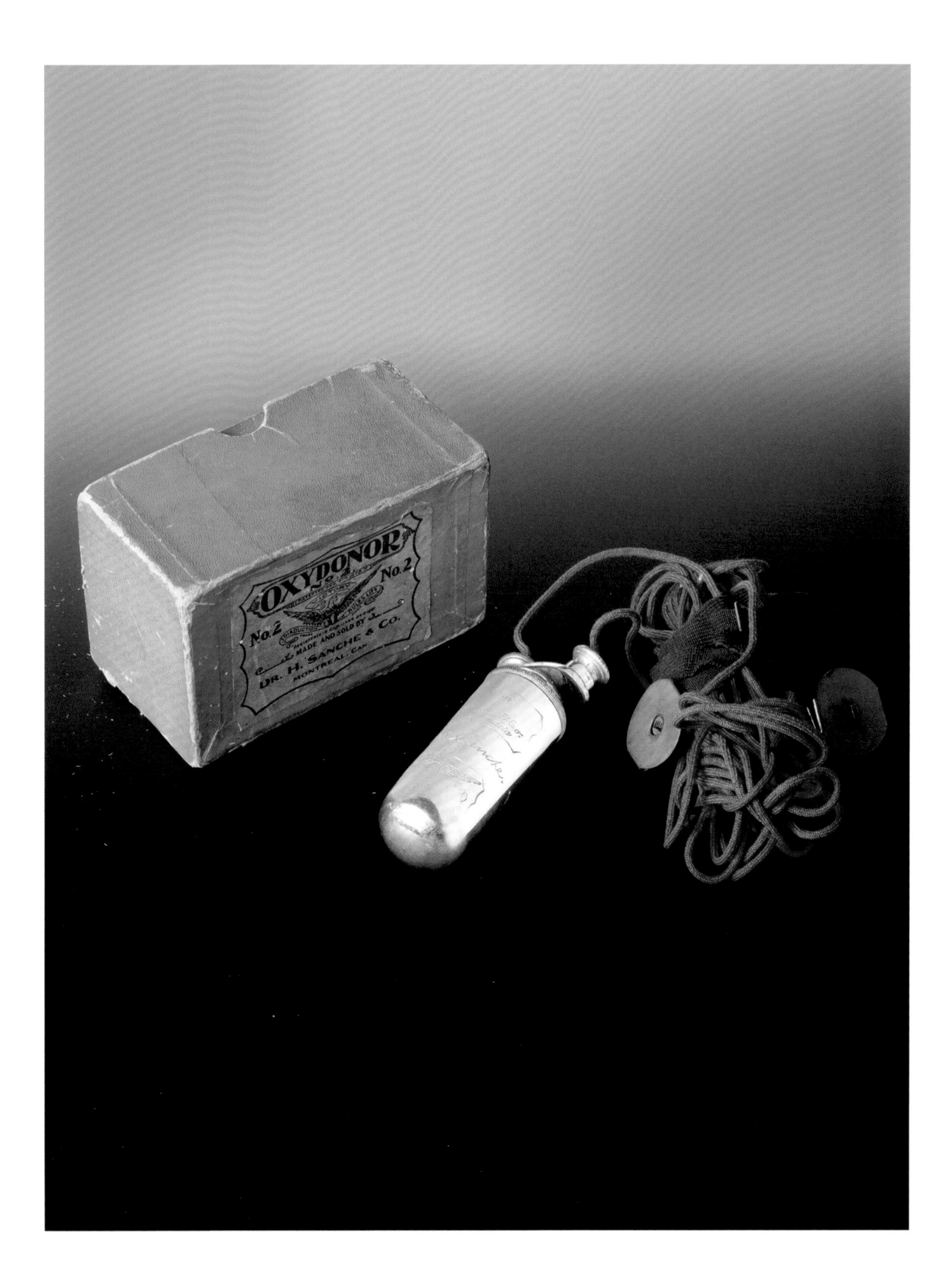
OXYDONOR
No.2
No.2
MADE AND SOLD BY
DR. H. SANCHE & CO.
MONTREAL, CAN.

THE OXYDONOR

Dr Hercules Sanche was a very successful quack, who claimed his "Oxydonor" machine would cure any and all ailments.

A device claiming to restore your body to health for practically any condition, without the need for a doctor, sounds appealing. It also sounds too good to be true. Such was the case of the Oxydonor, one of a range of cure-all devices known as "gas pipes". Invented in the 1890s by the notorious American quack Dr Hercules Sanche, many thousands of these devices were sold by mail order in the United States, Canada and Europe. Sanche claimed the Oxydonor could help the body cure a wide range of ailments, from pneumonia to fertility problems. Yet, even during its heyday, the Oxydonor was widely denounced as a fraudulent medical device.

Consisting of a metal cylinder connected to two wires, which were strapped to a patient's ankle or wrist, the Oxydonor's elaborate design convincingly resembled many of the electrical devices that were popular with nineteenth-century consumers. However, the Oxydonor claimed to harness the restorative powers of oxygen, not electricity, to help the body to treat itself. Purchasers were instructed to place the Oxydonor cylinder into cold water, and the device would filter oxygen from the water down the wires and into the patient to revitalize their body. With this pseudo-scientific rationale Sanche tapped into and exploited public awareness about, but limited understanding of, recent discoveries about the therapeutic benefits of oxygen.

By 1915 the Oxydonor was exposed as a fraud. Quack-busting investigators opened up the Oxydonor cylinder and found it to be full of sand rather than any oxygen-inducing mechanism. As governments and medical associations began to crack down on medical fraudsters, many of Sanche's competitors were put in jail. Sanche evaded prosecution by fleeing to Canada, where he continued to profitably sell his devices by mail order.

While the Oxydonor was effectively useless as a therapy, considering the other medical treatments available at the time – including bloodletting – the Oxydonor's popular appeal is perhaps understandable. For people who wanted to manage their own health or who were in desperate need of a cure, a one-off purchase, which could be used repeatedly to treat many conditions, could be seen to be prudent use of money. Though ineffectual, at least the device was unlikely to give patients any serious side-effects. They might even benefit from some bed rest.

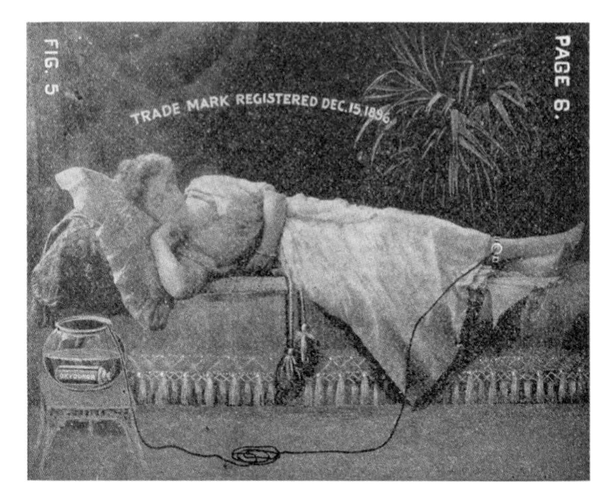

Opposite: The Oxydonor, 1898–1920.

Above: From the instruction manual for the Oxydonor, 1898–1920

HOSPITAL DE SANTA CRUZ

Cleverly crafted dioramas can give a miniature perspective view of historic interiors like this sixteenth-century hospital.

The number of dioramas on display at the Science Museum has declined greatly in recent years. These 3D scenes, rendered in a range of scales, have been considered old-fashioned by some critics and historically inaccurate by others. But even in a digital world, dioramas can still be intriguing and evocative forms of interpretation, at times providing glimpses into imagined, or reimagined, worlds, or recreating particular moments in history that have long since passed.

This simple diorama depicts the interior of the Hospital de Santa Cruz in the Spanish city of Toledo, at some time during the 1500s. But take away the four-poster beds and you could be looking into the interior of a church – a visual impression that says much about the hospitals of the time. And while the view it presents can be seen as a precursor of the hospital wards we are familiar with today, in many ways these were very different institutions. Some of them did offer forms of medical treatment, but many acted mainly as hostels for religious pilgrims and other travellers. They could house the sick, but might also assist the elderly, orphans and the homeless, or those unable to work. Poverty was perhaps the most common characteristic among these different "patients".

Across Europe many of these places of care were founded by leading public figures and wealthy elites, and nearly all were closely linked to religious organizations. This latter association is evident in their architecture. Hospitals were often built on land adjacent to churches or monasteries and could appear strikingly similar to religious buildings; they might even be one and the same building.

The Hospital de Santa Cruz was initiated by Cardinal Pedro González de Mendoza, primarily to assist orphaned and abandoned children in the city. Mendoza epitomized the associations between church, royalty and patronage, that were behind much of this earlier hospital provision. He was born into a powerful family and spent a career as a soldier, statesman and religious leader, fiercely loyal to the Spanish royal family throughout. Although he died while the hospital was still being planned, he left the project with royal support and enough money in his will to ensure it was completed some years later.

Above: Cardinal Pedro González de Mendoza, early benefactor of the hospital, portrait by Matías Moreno González (1840–1906).

Opposite: Diorama showing Hospital de Santa Cruz in the sixteenth century, made in 1901–70.

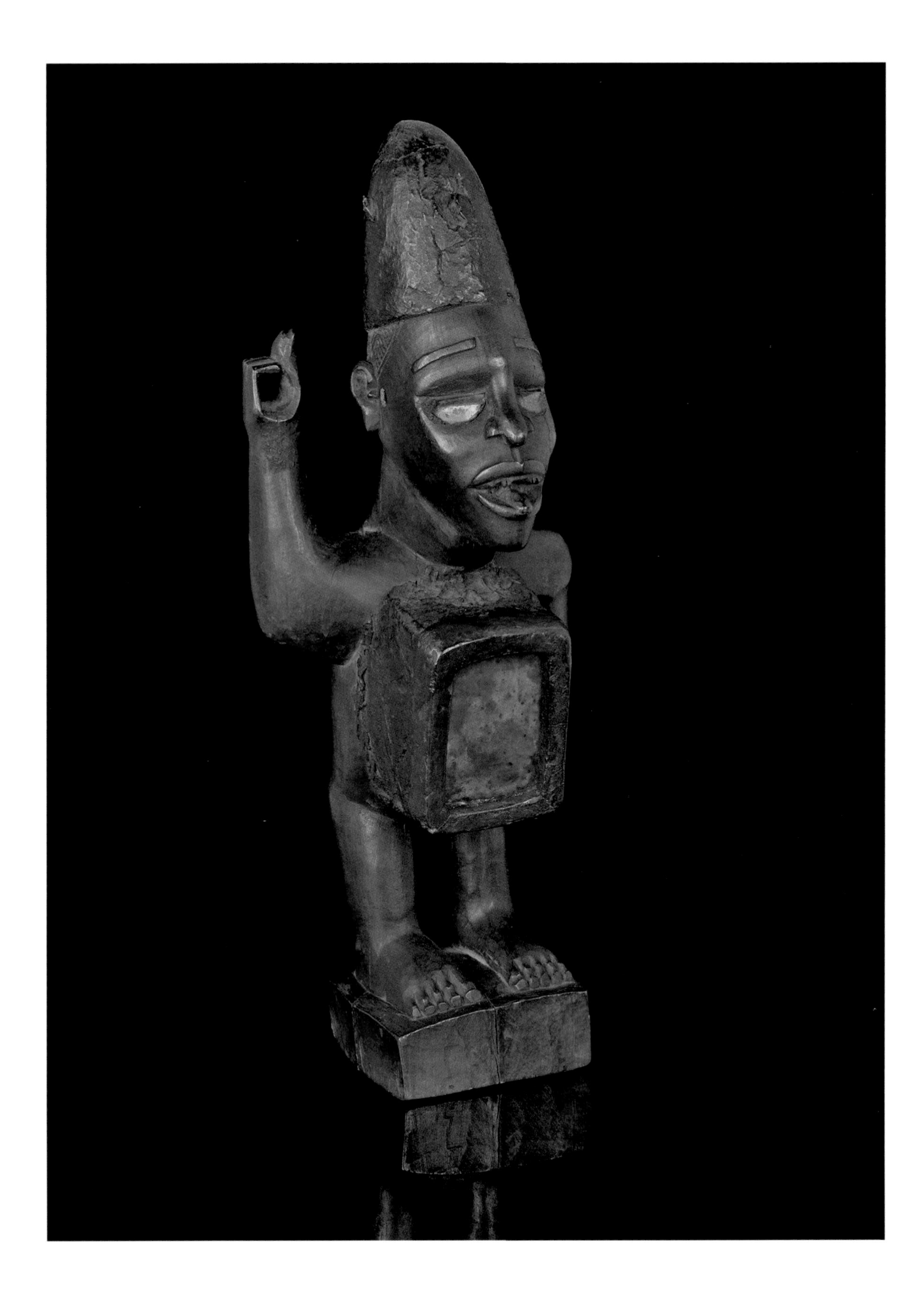

HUMAN-SHAPED CONTAINER FOR NKISI FORCE

The Bakongo people of West-Central Africa used these power figures, believed to contain nkisi spirits to fight illness.

These ferocious-looking wooden figures can appear somewhat intimidating, which is apt considering they were created to fight illness. Used by the Bakongo people of western Central Africa, they would have been used to contain a nkisi force. *Nkisi* roughly translates as "spirit" and is perceived as a power arising from the invisible world of the dead. To influence both the human and spirit world, for good or bad, this force must be persuaded to inhabit a contained space and to do the bidding of the person who activates it.

Containers usually take the shape of a statue in human form, and in English these are sometimes known as "Power Figures". These wooden figures were carved by a ritual specialist known as a *nganga*; the ones pictured here were created between 1880 and 1920. A cavity in the abdomen would be filled with medicinal herbs and other elements, called *bilonga*, which were believed to be beneficial in curing physical illness or alleviating social ills. These ingredients were extremely important, allowing the container to be inhabited by a nkisi force, and would have been stored either inside the head of the statue, or in a square pack in the abdomen, which is sealed with a mirror. Having invited a spirit or ancestor to inhabit this vessel, the nganga would then manipulate the nkisi to intervene in the human world on behalf of their clients: for example, to heal those suffering from illness and injury, or to seek out and punish evil doers at the heart of these misfortunes.

But why is the second container studded with nails? Different types of nkisi figure were created and used for a multitude of purposes and in a variety of ways. Known as a *nkondi*, translating as "hunter", second figure was used to find and punish those responsible for misfortune or injury. To activate these powers, a nail would be hammered into its body or head. The first figure is bare, which could mean that it was used for more benevolent purposes. However, it is standing in the hunter position, so perhaps this figure was simply collected immediately after creation, before it could be used.

Devoid of the medicine perceived to bestow potency on these figures, they would both now be regarded as empty vessels. Nonetheless, they literally embody the hopes and grievances of the individuals who used them, or their desire to control the unknown.

Opposite: Container for Nkisi force, Bakongo people, western Central Africa, 1890–1910.

Above; Nkisi figure, complete with nails, used to fix oaths and heal the sick, 1880–1920.

VOLVELLE TO CHART THE ZODIAC

Medical practitioners used this moveable, brass wheel to calculate the position of the moon within a patient's zodiac chart.

This volvelle, a moveable wheel chart, helped medical practitioners calculate the age, phase and position of the Moon within the zodiac. Each of the zodiac signs was thought to have a special dominion over specific parts of the body and, when the Moon was in particular star signs, it was an unfavourable time to perform treatments on the corresponding body part. This relationship between the star signs and the body is represented on the reverse of the volvelle. From antiquity to the eighteenth century, medical theory held that the human body was intimately connected to the surrounding universe. The four elements that made up the world (earth, air, water and fire) were also thought to make up the human body. Likewise, the four humours (phlegm, blood, yellow bile and black bile) were constantly changing in the greater universe as well as the individual body.

The names of the months and personifications of the zodiac signs are rendered around the edges of the volvelle. To use it, the missing lower dial and the lunar dial above would be moved to the date of the planned medical treatment. Together the two dials displayed the age of the Moon and illustrated which zodiac sign the Moon was in, or would be in, at the chosen moment.

The reverse of the volvelle is also richly illustrated. On the left of the central pillar is a Zodiac Man, a representation of the celestial influences on the human body. The twelve zodiac signs are engraved on top of the naked male form, including Gemini over the arms, Pisces on the feet and Scorpio over the genitals. To the right of the pillar there is a figure of a medical practitioner holding a cupping horn, used for bloodletting, and a drinking vessel, perhaps with medicine. This figure is accompanied by an unusual representation of the ancient author Claudius Ptolemy. Due to confusion with the Egyptian Ptolemaic dynasty, Claudius Ptolemy was often rendered as a king and here we see him wearing early modern armour. The numbers and letters around the outer edges of the brass disc can be used to calculate the date of Easter and other movable feasts.

Although there are many paper and parchment volvelles hidden away in manuscripts and books, metal ones are much rarer and would have functioned not just to aid in medical practice, but also as an emblem of the user's medical status and scholarly authority.

Opposite: Brass volvelle used to calculate ideal times for medical treatments such as bloodletting, 1583.

Left: The reverse shows a richly illustrated scene.

LES·IOVRS
DE·LAN
LA·LVNE
IANVARIVS
FEBRVARIVS
MARTIVS
APRILIS
MAIVS
IVNIVS
IVLIVS
SEPTEMBER
OCTOBER
NOVEMBER
DECEMBER

VOTIVE MALE TORSO

Roman society offered their Gods votives of body parts made of bronze or clay as part of a prayer for health.

Is it a sculpture? An anatomical model? This almost life-sized terracotta torso is actually a 2,000-year-old Roman votive offering – a gift for the gods, to request or give thanks for a safe recovery from illness or accident. The teardrop-shaped opening from the breast bone to the upper abdomen reveals a stylized depiction of the internal organs, possibly gleaned from animal carcasses since human dissection was considered taboo in Roman society. Still visible on the trunk and limbs are traces of the vibrant red paint used to decorate it.

The Science Museum cares for more than 800 of these "anatomical votives", ranging from skilfully cast bronze miniatures to strikingly lifelike limbs and organs, mass-produced from local clay using moulds. Archaeologists believe that they were deposited by pilgrims visiting ancient Greek and Roman sanctuaries, to express gratitude for healing or to draw the gods' attention to an afflicted area of the body. Seeing them gathered together in vast quantities must have been a powerful and humbling experience, reminding worshippers that they were far from alone in their suffering and testifying to others' eventual recoveries.

With little to go on but the objects themselves, the specific reasons behind their deposition are open to debate. Few exhibit visible signs of disease, and certain types – such as eyes and ears – may have conveyed symbolic as well as literal meanings. While some indicated problems with sight and hearing, others were perhaps intended as pleas for the deities in question to see the dedicants' suffering and listen to their prayers. Similarly, legs and feet may have been offered in gratitude for the successful completion of a long journey rather than in relation to a specific injury or disorder. Full torsos like this one, reputedly excavated from Isola Farnese on the outskirts of Rome, were exclusive to the ancient Italic civilizations. A votive of this size would have been a considerable investment. The person who gifted it possibly suffered from digestive problems or was experiencing more generalized symptoms.

Opposite: Votive male torso, dissected to show viscera, reputedly from Isola Farnese, Roman, 200 BCE–200 CE.

Above: Votive heads in the Science Museum's store.

The tradition of votive giving still exists today in many cultures and faiths. Votives are found decorating altars in Christian churches across the world, especially in southern Europe and other places with a strong Catholic tradition. Although modern votives are more commonly made of sheet metal or wax, the intention behind their use remains much same as in the ancient world. The desire for good health is indeed timeless.

WATER FROM THE MEDICINAL WELL AT EPIDAURUS, GREECE

Many global religions include a belief that drinking or bathing in holy water has strong healing outcomes.

Using a glass Evian bottle he had at his disposal, Captain Peter Johnston-Saint collected water from the well at the Temple of Asklepios, Epidaurus. Deserted since *c.*700 CE, the ruins became a popular visitor spot in the late-nineteenth and early-twentieth centuries; in 1930, Johnston-Saint was only one of many tourists. Johnston-Saint was interested in the temple at Epidaurus because of its connection to Asklepios, the Greco-Roman god of healing; he noted that Epidaurus was the main temple to Asklepios, to which all others could trace their origins. At temples dedicated to Asklepios those experiencing illness underwent a process known as incubation. While you slept, it was believed you would be visited and cured by Asklepios himself, or by his daughters Panacea and Hygeia. If you had dream visions, a priest of Asklepios could interpret them and suggest a treatment. Before visitors slept, they would wash in spring water nearby. Although this practice had died out by Johnston-Saint's time, he noted that local people still visited the spring to help with rheumatism.

Above: Model of the Asklepion at Epidaurus, Greece, 1936, made after Johnston-Saint's visit to the site.

Opposite: Bottle, originally for Evian water, containing water from the medicinal well of the temple of Asklepios at Epidaurus, Greece, collected by Captain Peter Johnston-Saint in 1930.

Johnston-Saint was one of the many agents who built up Henry Wellcome's medical collection, which is now on long-term loan from the Wellcome Trust to the Science Museum. He meticulously recorded his movements for his employer, Henry Wellcome, who had a voracious appetite for collecting. While medicine collections conjure images of anatomical specimens and surgical instruments, Wellcome's ambition was to collect the whole experience of life and death, health and illness, including religious beliefs from across the globe. Wellcome was a pharmaceutical magnate who co-founded Burroughs, Wellcome & Co, in 1888 and established the Wellcome Historical Medical Museum to display his burgeoning collecting. Johnston-Saint travelled thousands of miles on behalf of Wellcome before taking up a permanent London-based position, caring for the Wellcome Historical Medical Museum until his retirement in 1947 after 27 years of service.

Johnston-Saint collected waters from all over Europe, including the Holy Land and Lourdes in France. Many of his samples were bought as souvenirs, just as you can do today at sites of pilgrimage across the world or even online. While the water Johnston-Saint collected in 1930 may not be exactly the same water that ebbed and flowed when ancient Greeks and Romans visited their site of pilgrimage, the memories of those people's beliefs are captured in this bottle.

WATER FROM THE
MEDICINAL WELL
AT THE TEMPLE
OF AESCULAPIUS
AT EPIDAMOS
Brought by Capt.

14509.

NETSUKE SHOWING THE HARE IN THE MOON

Netsuke are Japanese ornaments carefully carved from one piece of ivory, often representing images relating to belief.

Opposite: A netsuke in the form of a rabbit-like creature using a pestle and mortar, Japan, 1701–1900.

Left: Netsuke in the Science Museum's collection.

Wearing a kimono, this rabbit-like creature is grinding ingredients in an oversized pestle and mortar. Measuring just under 4 cm (1½ inches) and carved from a single piece of ivory, the decorative piece is known as a *netsuke*. The word comes from the Japanese, "ne" meaning root and "tsuke" meaning "to attach". Netsuke, pronounced "net-skey", are used to hang objects called *sagemono*, such as medicine boxes or tobacco pouches, from the sash, or obi, of a kimono – a traditional form of Japanese dress. Each netsuke has two small holes carved in the base for this purpose.

More than 200 netsuke are held in the Science Museum's medicine collection, taking a variety of forms from skeletons and skulls to animals, deities and people. Each one is a visual feast, and it takes several viewings to appreciate the exquisite detail in each. Netsuke were fashionable in Japan from the seventeenth century onward, and by the mid-nineteenth century they had captured the imagination of European collectors. In the context of the Science Museum, a collection of netsuke might feel at odds with an institution that celebrates science, technology, engineering and medicine. In fact, the medicine collections have their roots in businessman collector Henry Wellcome's ambition to collect the global experience of health and illness, life and death. While Wellcome's vision was unprecedented, he was not alone in collecting netsuke, either in Japan or at auction – most national museums in the United Kingdom have a collection of these ornaments. It is possible though that this example was collected for its visualization of a pestle and mortar, and for its connection to pharmaceutical manufacturing, which is how Wellcome made his fortune.

Another potential reason is the representation of belief, another part of the Science Museum's medicine collection. Hares or rabbits are part of the twelve animals of the Japanese zodiac, and Asian, African and European traditions all have stories associated with the hare or rabbit as a lunar messenger. It is thought that the imagery of this netsuke refers to the "hare in the Moon". The creature mixes the elixir of immortality with his mortar and pestle.

This netsuke is a curatorial favourite, for its artistry and for the multiple stories that it evokes. While we do not know the maker's original intentions, each viewer of this miniature carving brings their own history to the piece to make their own story – an experience that continues to inspire.

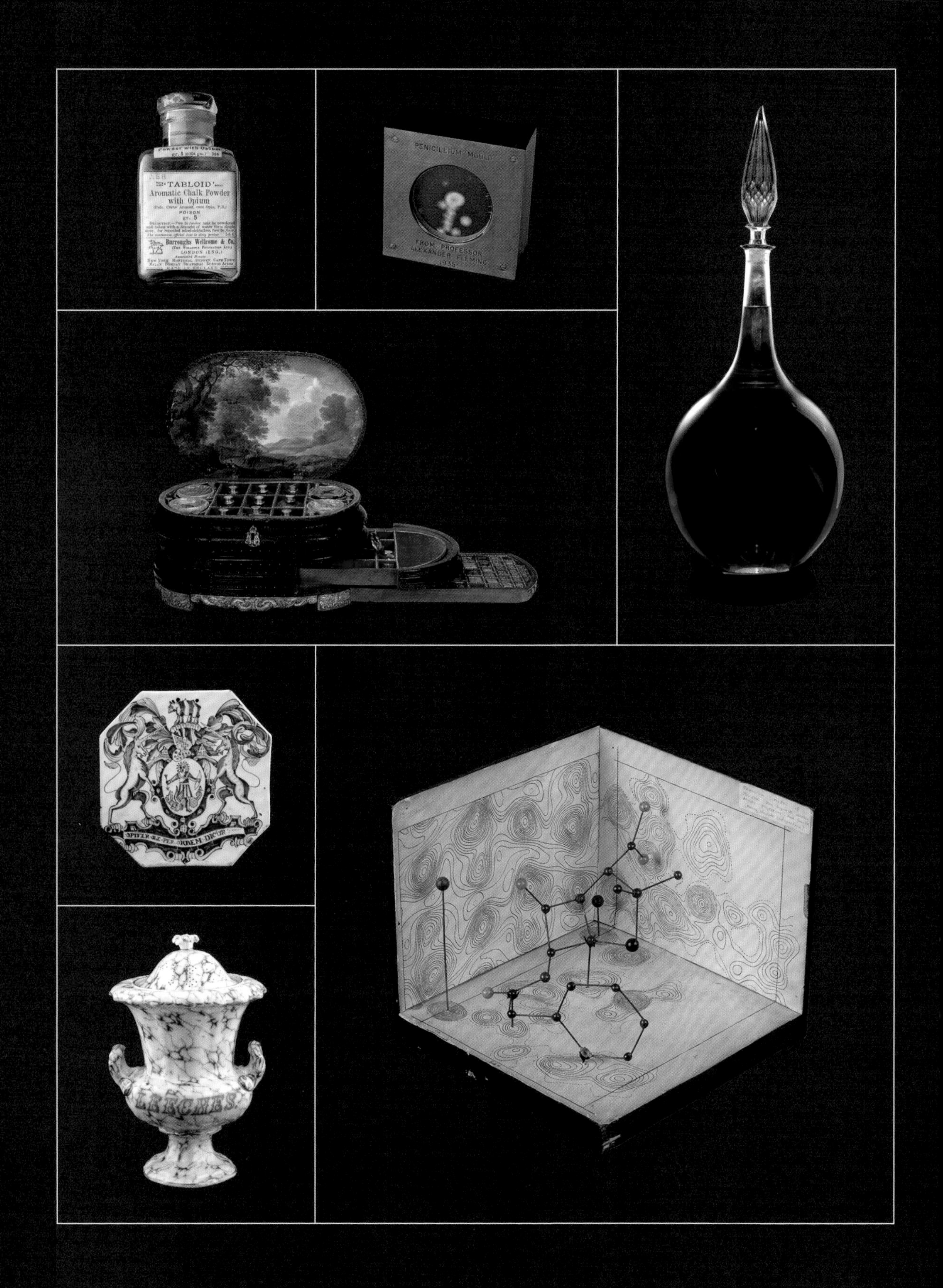
'TABLOID'
Aromatic Chalk Powder
with Opium
POISON
gr. 5
Burroughs Wellcome & Co.
LONDON (ENG.)
PENICILLIUM MOULD
FROM PROFESSOR
ALEXANDER FLEMING
1935
LEECHES

8

DRUGS AND PHARMACY

The pills and potions that can cure or ease our ailments are in some ways at the heart of our understanding of medicine. Plant- or animal-based products, artificially created chemicals, or completely natural ingredients can all cause reactions in the human body, ranging from the mild to the toxic. The medicine cabinet that stands in every household bathroom has a complex identity and history. The control of pain, infection and fertility has been of enormous benefit to humanity, but the misuse of drugs has always accompanied the beneficial effects, and the addictive nature of many medicines can have fatal effects. Drugs also form the cornerstone of the commercial side of medicine, from the historic apothecary selling his wares in his own shop to the huge pharmaceutical companies responsible for researching, developing and selling drugs on the high street today.

THE PHARMACIST'S CARBOY

Hard-working apothecaries filled glass carboys with coloured water to attract the attention of passers-by.

How would you find a pharmacy if you couldn't read? Carboys were glass containers filled with brightly coloured liquids and placed in the windows of pharmacies to indicate to passers-by that this was somewhere they could buy medicines. They were synonymous with pharmacies in English-speaking countries between the seventeenth and twentieth centuries.

The pharmacy window was used as a form of advertising; behind the carboys, the windows were stuffed with all the products that the pharmacy stocked. Inside, the walls were lined with jars filled with various herbs, spices, powders and pills. Liquid medicines were stored in glass bottles, some of which were ridged to indicate that the contents were poisonous. Much like carboys, a lot of pharmacy-ware had a decorative as well as a practical function.

Unlike today, pharmacists and apothecaries (in many ways the precursor to the modern general practitioner) had a difficult relationship with one another. Apothecaries charged patients for medical advice, then prescribed and sold medicines to patients directly, bypassing the need for the pharmacist. Pharmacies, however, provided free medical advice with every purchase, so they competed with apothecaries for custom. Once the Worshipful Society of Apothecaries was created in 1617, apothecaries had to undertake an apprenticeship in order to gain accreditation. Over time they moved away from selling medicines, electing instead to prescribe them and to diagnose their patients. With the gap between prescribing and dispensing growing, the modern-day pharmacy emerged.

Unlike apothecaries, pharmacists were not officially trained or apprenticed until 1841, when their governing body, the Pharmaceutical Society, was established. For years pharmacists struggled to shake off their "quack" reputation, which they had attracted through selling a variety of non-medicinal products. The formation of

the Pharmaceutical Society legitimized the profession, creating a national membership and a standard qualification for all pharmacists, giving credibility to their advice and challenging their poor reputation. Subsequently the Pharmacy Act 1868 limited the selling of medicines and poisons to qualified pharmacists and druggists only. Eventually pharmacists and GPs developed very distinct remits, and now they work together to provide frontline healthcare to the British population.

This object dates from the early nineteenth century but was purchased by the Wellcome Foundation in 1962. It formed part of a larger acquisition that included the contents and the shop fittings from Tyler's Pharmacy in Kensington, London, owned previously by Norman Fielder Tyler. Though rarely used today, carboys are an example of early trademarking and advertising to an illiterate public.

Opposite: A caricature of a pharmacist, his apprentice and a customer, with the brightly coloured carboys in the shop window, 1825.

Right: Carboy from a London pharmacy, 1840–90.

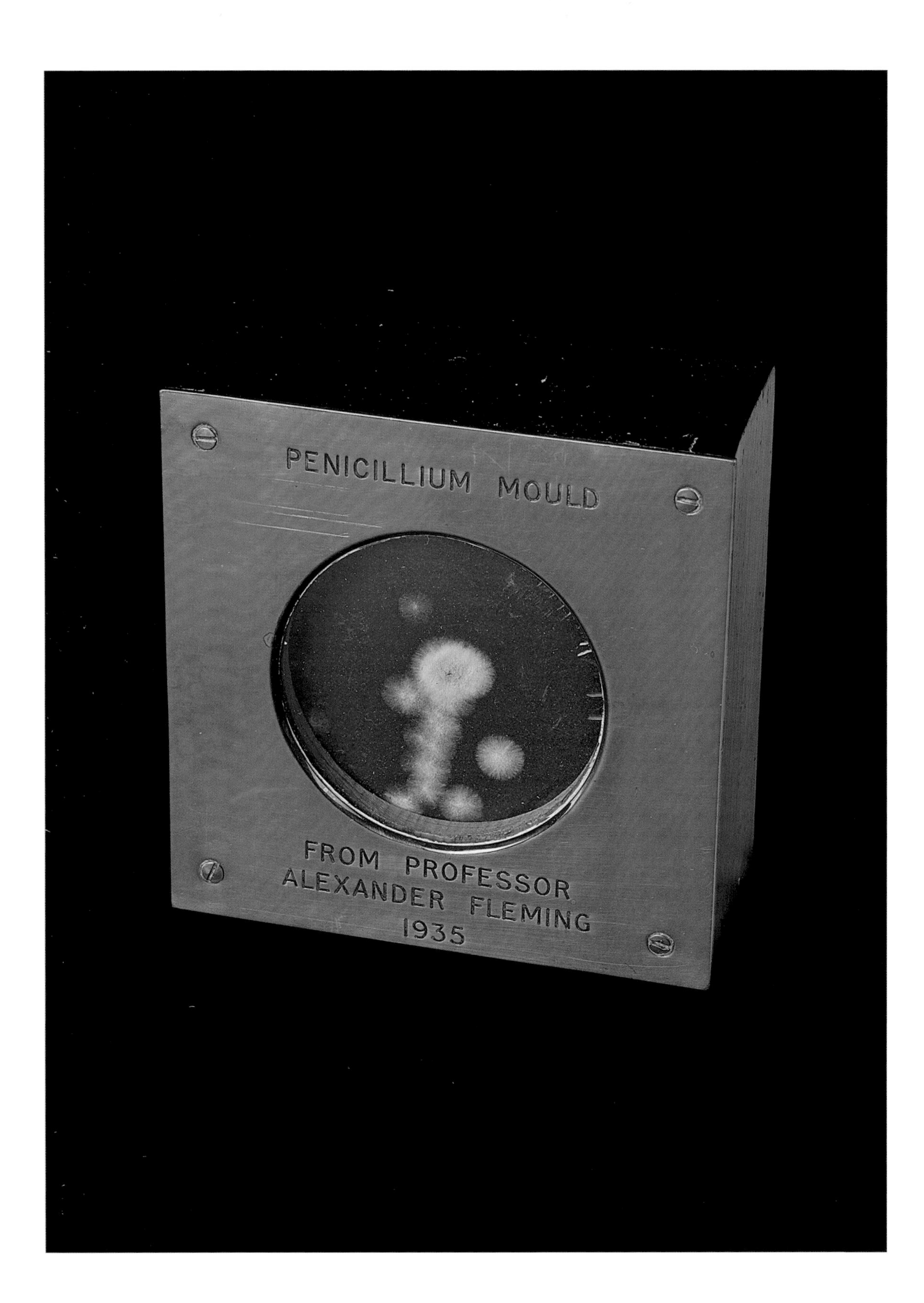
PENICILLIUM MOULD
FROM PROFESSOR
ALEXANDER FLEMING
1935

PENICILLIUM MOULD

Alexander Fleming made a gift of a sample of *Penicillium* mould to his friend Douglas Macleod.

Penicillin began in as a laboratory curiosity in 1928. Fifteen years later it was hailed as a wonder drug, heralding the use of antibiotics as a way of combating infection. In between there was much research but also rewriting of its story. This sample documents a critical transition in the medical world's perception of penicillin's potential.

Alexander Fleming, in a laboratory at St Mary's Hospital in Paddington, first observed that the liquid exuded by a *Penicillium* mould affected bacteria. He put some work into invetigating the potential significance of his observation. At the time, however, it seemed that the true future of medicine lay in boosting the body's immune system, not in new medicines. Fleming's own boss was a leading campaigner for such an approach. Wonder drugs were widely dismissed as the business of quacks.

In 1935, after much hesitation, the German company Bayer announced its medicine Prontosil, the first "sulpha" drug, and in October the company's medical director made a presentation in London. He announced a first real success in the treatment of bacterial diseases, opening the prospect of others. The lecture was attended by Alexander Fleming, who had brought his friend and colleague, the gynaecologist Douglas Macleod. At the end of the lecture Fleming turned to his friend and said: "I've got something much better than Prontosil but no one will listen to me", as reported in André Maurois's *The Life of Sir Alexander Fleming*. Fleming subsequently gave Macleod some spores from his *Penicillium* on a watch glass. The family later had the glass and its precious sample mounted in a brass holder, and at the end of the twentieth century this was purchased by the Science Museum.

Fleming himself did little further research on penicillin, but the success of Prontosil had certainly changed attitudes to drug treatment. In 1939 Prontosil's discoverer was awarded the Nobel Prize for Medicine. By then several other new products and avenues of research were being followed. Penicillin was explored again at Oxford and although this work began as fundamental science, it quickly became a search for the new wonder drug. That vision had been presaged by Fleming's experience in 1935 and the gift of this now precious sample.

Opposite: Sample of *Penicillium* mould presented by Alexander Fleming to Douglas Macleod, 1935.

Below: Box of free sample Prontosil ampoules, by Bayer, Germany, 1936–40.

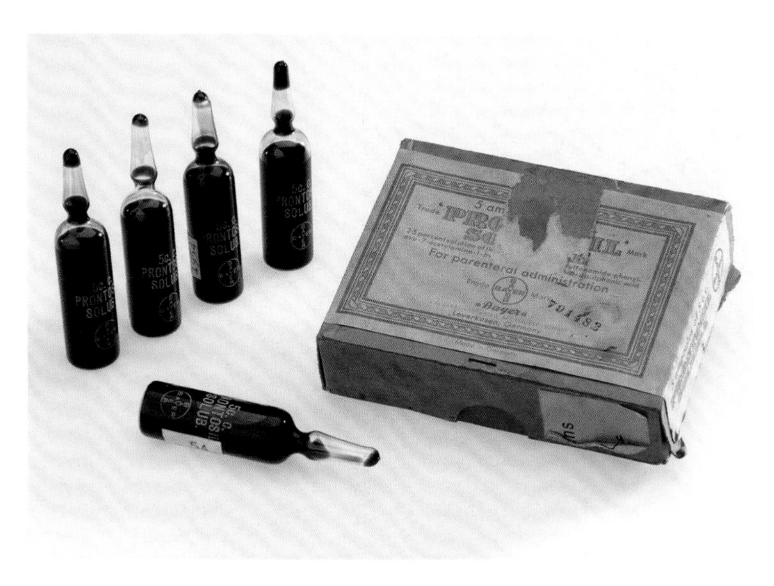

TRACY – THE TRANSGENIC EWE

A landmark in animal-human hybrids, Tracy the Sheep was genetically modified to produce a human protein in her milk.

Once the stuff of myth and science fiction, today human–animal hybrids are a reality. Tracy was a transgenic ewe, genetically modified by injecting human DNA into a sheep embryo to produce alpha-1-antitrypsin in her milk. This human protein controls the enzymes that cause lung damage and was thought to offer hope in the treatment of cystic fibrosis and emphysema. Created at the Roslin Institute in Scotland, the same place that later cloned Dolly the Sheep, Tracy was born in 1990 and lived for seven years. During her lifetime she passed her genes on to her offspring, who also carried this curious adaptation. The hope was that her milk and that of her kin would supply material for advancing medicine and curing disease. When she died, Tracy was acquired by the Science Museum and taxidermied in memorial to this moment in the history of medicine.

Once viewed as radical, the genetic engineering of animals in pursuit of medicine is now relatively widespread, and biotechnology is an established industry in its own right. Humans have produced the high-strength fibre BioSteel™ from goats with spider silk protein in their milk. Other transgenic animals include creatures that glow florescent green under ultraviolet light, mice predisposed to cancer as models for research, and even pigs with organs more resistant to being rejected by the human body in transplant surgery.

While in some instances such "pharming" practices can be used to increase understanding of human health and offer hope for combatting disease, the ethical debates surrounding this manipulation of life are multiple. Chief among these relate to animal welfare and the exploitation of non-human animal life for human ends. Should humans meddle with the bodies of other animals to serve their own needs? Does the advancement of medicine warrant this utilitarian use of animal bodies? And what sort of indeterminable identities are created when human and non-human animal bodies become entangled in these ways?

There are no straightforward answers to these questions, and ethical debates will continue to rage. Yet one thing is certain: where these practices continue – and in instances where animals are used for the purposes of science and medicine more broadly – they undoubtedly need to be approached with care, respect and a sense of responsibility.

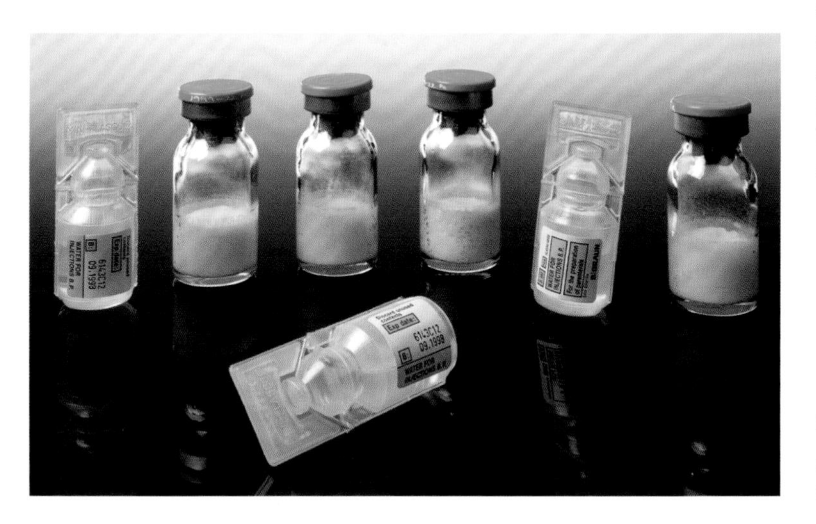

Left: Samples of alpha-1-antitrypsin, 1998.

Opposite: Tracy, a transgenic ewe, preserved through taxidermy in 1997.

THE FIRST ORAL CONTRACEPTIVE PILL

The launch of the first contraceptive pill created unprecedented freedom for women and allowed them to control their own fertility.

The oral contraceptive tablet, better known simply as the Pill, is part of daily life for many women in Western society, and yet the impact of this tiny pill has been monumental.

Although birth control in one form or another has been practised for centuries in the UK, it wasn't until 1921, with the introduction of the first birth control clinic, that women were able to seek contraceptive advice. Previously many women had had to resort to unreliable and sometimes dangerous methods of terminating pregnancy. The first discussions about the development of a contraceptive pill occurred in the US after the Great Depression of the 1930s. With widespread poverty sweeping the nation, it was hoped that the Pill would result in smaller families and greater resources for each child.

On 9 May 1960, the US Food and Drug Administration (FDA) approved Enovid, the world's first birth control pill. Though there were many contributors to the final product, Enovid's contraceptive properties were proven by biologist Dr Gregory Pincus, who showed that a combination of oestrogen and progestin suppresses ovulation. Launched the following year in the UK, the Pill was initially only available to married women, but this law was relaxed in 1967 and by 1974 all contraceptive advice and prescriptions were free of charge on the National Health Service, irrespective of age or marital status.

This revolutionary tablet empowered women to take charge of their own fertility, dramatically contributing to women's liberation and to the sexual freedom of the so-called Swinging Sixties. Prior to the introduction of the pill, many women's rights may have existed in the Western world, but with the unpredictable arrival of children the reality was quite different. Women could now choose when to have children, no longer constrained by the reproductive consequences of sex.

Not everyone approved of the Pill. Many suggested it encouraged promiscuity and some religious groups even labelled it a form of abortion. Long-term use of the Pill has also been associated with increased risk of developing conditions such as thrombosis, cancer and depression, and yet oral contraceptives remain among the most commonly prescribed drugs worldwide. Enovid itself was discontinued in 1988. Decades after the Pill's introduction, countless improvements to their safety have been made and, though this tiny pill still attracts strong and conflicting opinions, it is impossible to deny the impact it has had and continues to have on women's lives.

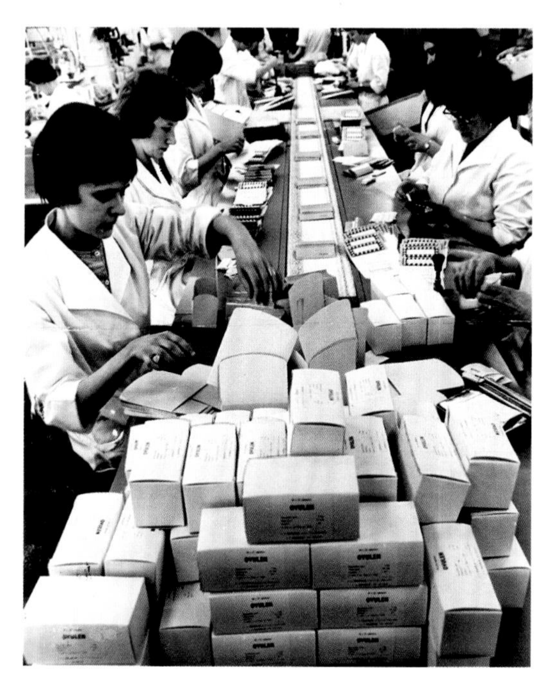

Above: Women packing boxes of contraceptive pills, 1965.

Opposite: Enovid, G.D. Searle and Company, early 1960s.

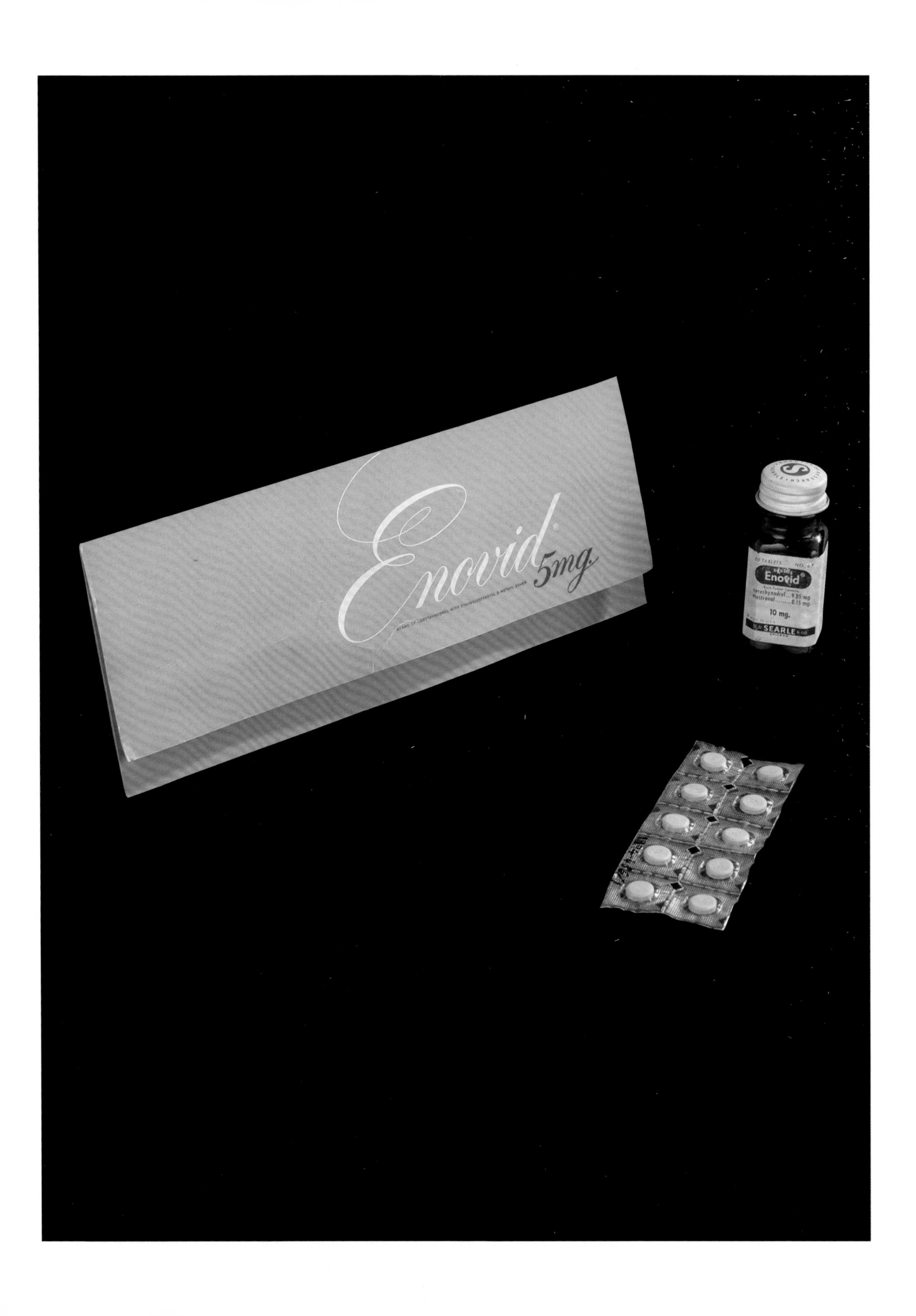
Enovid
5mg.
Enovid
9.85 mg
0.15 mg
10 mg.
SEARLE

THE GIUSTINIANI FAMILY MEDICINE CHEST

Although most people own an ordinary medicine cabinet, the chest belonging to the Giustiniani family is a work of art.

For the Giustiniani family, this luxurious chest was a first line of defence, much like our own first aid kits at home. The contents of this ornate medicine chest give a tantalizing glimpse into the medical concerns of one Italian family in the sixteenth century. Vincenzo Giustiniani was the last Genoese governor of the island of Chios in the eastern Aegean Sea. He ruled Chios from 1562 until the Turkish expelled the Genoese in 1566, after an occupation of over 200 years. The chest remained in the family until it was purchased in 1924 by an agent working for Henry Wellcome, who was attempting to build a collection that encompassed the whole experience of human health. The chest has remained one of the stars of the Science Museum's medical collections since some of Wellcome's vast holdings were transferred there on loan in the late 1970s.

Wealthy enough to afford such a luxurious and unique item, the family probably had their own doctor on staff. A miniature pestle and mortar with small weighing scales suggests that individual doses could be made as required. Bespoke made, each of the 127 bottles and boxes has its own space in the three drawers of this bespoke medicine chest. Some of the contents remain a mystery, either their labels obscured, or the containers completely emptied. Meticulously handwritten labels in Italian describe treatments that seem both familiar and fantastical to us. Prepared treatments for wounds, teeth and piles lie within, as well as individual ingredients such as mustard oil, cinnamon, rhubarb and unicorn horn.

"Unicorn horn" was usually the long tooth of a male narwhal, a species of whale, ground into a powder. It was believed to have medicinal qualities. Rhubarb, more familiar now as a food, was popular for stomach, lung and liver problems. Trading with the Spanish, who held territories in the Americas, the Giustiniani family could also take advantage of new medicines and plants arriving from the so-called New World. Mexican mechoacan root, commonly known as jalap, was believed to purge excess fluids in the body, while ground guaiacum wood was a prized treatment for syphilis.

Opposite: The Giustiniani family medicine chest, 1562–66.

Below: Selection of contents from the medicine chest includes a tiny pewter box for unicorn horn, a pestle and mortar and a set of scales.

SOCIETY OF APOTHECARIES' PILL TILE

Apothecaries mixed medicines with honey or liquorice and then rolled and cut pills on earthenware tiles to make dosage as accurate as possible.

What part does a tile play in the pill production process? Apothecaries used tiles like this one to prepare pills for their customers. They mixed the various parts that made up the medicine together, then combined it with a sugar or liquorice solution to bind it into a thick, pliable substance. After that it was rolled into a long sausage shape on the pill tile, before being cut into the right size and shape for individual doses. While today our pills are standardized in form and medicinal content, it was difficult to regulate handmade pills, so the dosage was often unreliable.

This pill tile is an especially interesting example. It is adorned with the arms of the Worshipful Society of Apothecaries, signifying membership of the Society and the professional status of its owner. The Society was formed in 1617, so it is thought this object dates from the 1600s. The Society's arms depict Apollo, the Greek God of light, music and healing. He is surrounded by two unicorns and a rhinoceros, the horns of which are believed to have significant healing properties. The tile shows Apollo defeating pestilence, which is portrayed as a wyvern, a legendary creature with a dragon's head and wings and a reptilian body. Accompanying this image is a Latin motto, which translates to: "And I am called throughout the world the bringer of aid."

The apothecaries were originally part of the Worshipful Company of Grocers, which can be traced back to the Guild of Pepperers formed in 1180. While the profession had a rich and lengthy history, it was not until James I gave them their Royal Charter that they separated from the Grocers, distinguishing the selling of drugs from other merchandise. In 1704 apothecaries were given the right to practise medicine, and the Apothecaries Act of 1815 granted the Society the power to regulate and license medical practice in the United Kingdom.

The Society still exists and works as an examining body for specialized medical qualifications, while caring for the relics and documents relating to the regulation of British medical practitioners. Pill tiles became obsolete in medical practice after the development of new technologies that could mass-produce drugs with standardized shapes and dosage. Now they remain as testaments to the craft of pill making and the interior décor of pharmacies gone by.

Left: A print of an apothecary from 1830, made up of iconic pharmaceutical equipment, including a pill tile, medicine jars, knives and pill pots.

Opposite: Society of Apothecaries' pill tile, seventeenth century.

OPIFER QUE PER ORBEM DICOR

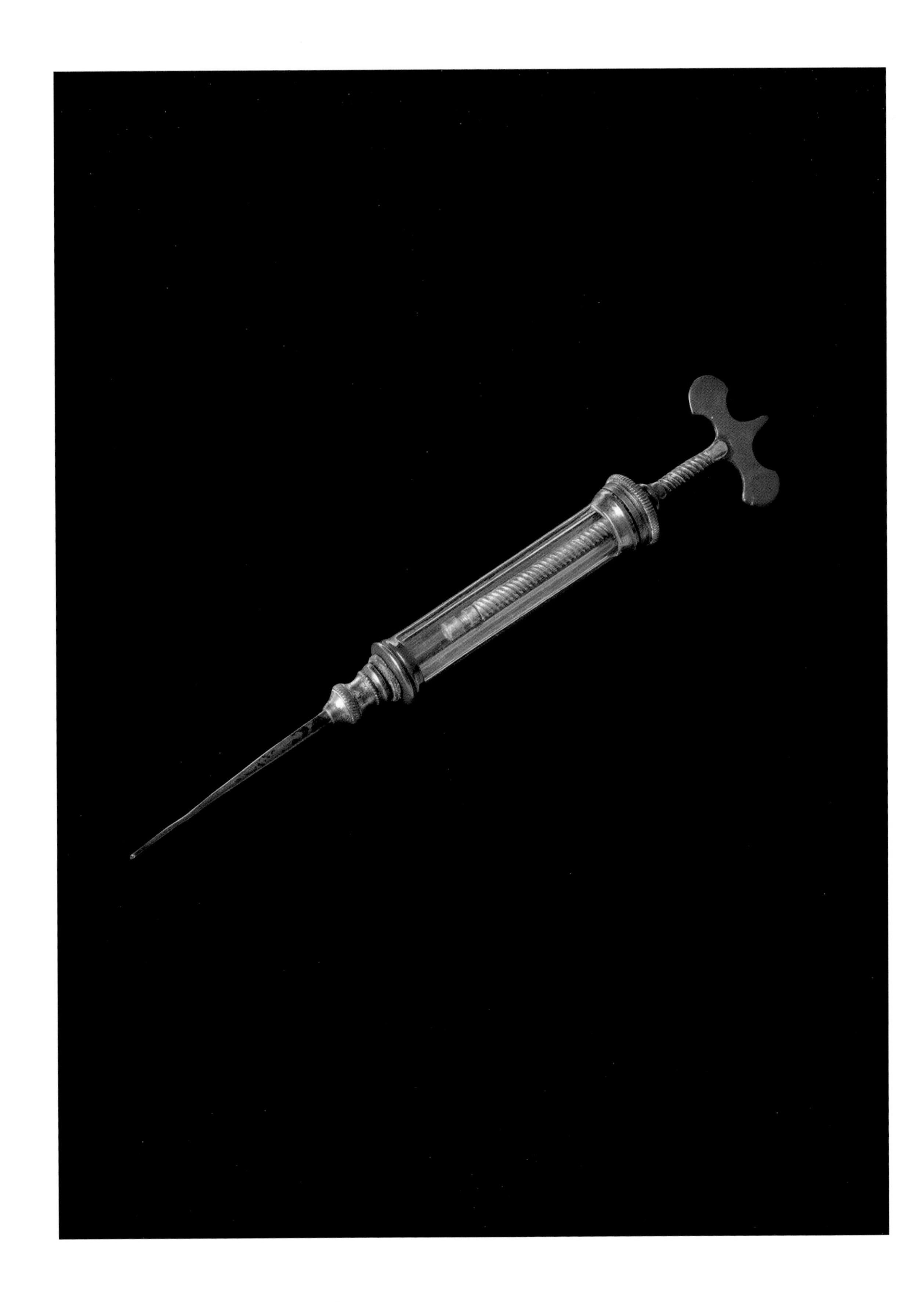

HYPODERMIC SYRINGE

Hypodermic syringes enable fast-acting drugs to be injected directly under the skin to ensure a rapid response.

What is your earliest memory of going to the doctor? It might be being jabbed with a sharp needle. Ouch! Hypodermic needles are used with a syringe to inject drugs or extract fluids from the body (hypodermic means "beneath the skin"). They are used in many of the most common and routine medical procedures - from donating blood to vaccinating children. They are also used to administer life-saving treatments.

The idea of injecting medicine into the body has its origins in ancient Greek and Roman medicine. It was not until the seventeenth century, however, that architect and anatomist Christopher Wren began to experiment with injecting drugs into dogs using a goose feather. These unsuccessful experiments put injection out of favour for decades. Then, in the 1850s, the French veterinary surgeon Charles Gabriel Pravaz and the Scottish doctor Alexander Wood developed a new kind of syringe with a hollow needle, which was fine enough to pierce the skin. Syringe barrels were initially made of metal, but by 1866 they were made from glass, enabling doctors to see how much medication remained in reserve. The needle and syringe shown here is made from silver with a glass barrel. Unlike modern syringes, which use a plunger, this one works by turning the screw at the top to inject the liquid. This would have been difficult and fiddly and required a skilled operator. The name Mathieu, a French surgical instrument maker, is punched into the syringe. At this time hypodermic needles were widely produced, but there were only a small number of injectable drugs available.

The twentieth century saw a transformation in the use of hypodermic syringes. An initial push came from the discovery of insulin in 1921, and its use in the treatment of diabetes. Later the Second World War encouraged the early development of partially disposable syringes for injecting penicillin on the battlefield. In 1956 the New Zealand pharmacist Colin Murdoch invented the disposable plastic syringe.

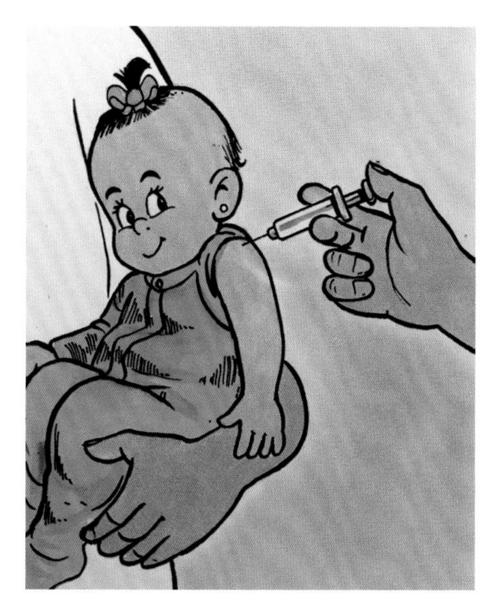

Opposite: Screw action hypodermic syringe, 1851–90.

Above: Childhood vaccinations are one of the many uses of hypodermic needles in modern medicine. This page is from a booklet entitled *Las Vacunas*, was used in public health campaigns in Colombia in the 1980s.

Hypodermic needles are ubiquitous in modern medicine, symbolizing the extraordinary array of medicines and vaccines available to prevent and treat illnesses. However, since their invention hypodermic needles have also been associated with contamination and drug addiction. Alexander Wood and his wife became addicted to injected morphine after Wood experimented with the pain-relief drug for treating neuralgia. Clearly, this medical marvel has a dark side.

MOLECULAR MODEL OF PENICILLIN

Dorothy Crowfoot Hodgkin dedicated most of her working life to working out the chemical structure of insulin and penicillin.

Almost a century after Alexander Fleming discovered that a juice from *Penicillium* mould killed bacteria, we still use penicillin. Chemists have modified the raw product produced by the mould to make it more widely effective and convenient to use. Some variants of penicillin can combat most of the bacteria resistant to more commonly used medicines, though doctors do increasingly encounter germs that have evolved to combat all antibiotics.

We are able to modify raw penicillin because its structure was decoded soon after it came into use. This was made possible by the skill of British scientists associated with the physicist John Desmond Bernal. They had become world leaders in using the reflections of high-energy X-rays from inside crystals to determine the structures of complex chemicals.

Dorothy Crowfoot Hodgkin at the University of Oxford had been a Bernal pupil and would devote much of her life to the structure of insulin, the chemical used to treat diabetes. But penicillin, a smaller molecule, proved a simpler challenge. During the Second World War she obtained some of the new product and sought its structure. Within a couple of years she was successful and made the model shown here. Her intense project had required new kinds of resource – not just theory and experiment but also lots of computing. Her grant officer at the Medical Research Council could not believe that the cost of buying time on huge electronic calculators, called tabulators, was actually greater than the cost of her laboratory work. Hodgkin's work was prompted by the hope that if the structure of penicillin were known, then the medicine could be entirely synthesized in the laboratory. In the event, making antibiotics synthetically rather than using moulds has generally not been economic.

Crowfoot Hodgkin would become world famous for her decoding of complex chemicals with X-ray crystallography. In addition to penicillin and insulin she was also well known for her work on vitamin B12. The work she and her colleagues conducted also inspired designs that were widely used in the 1951 Festival of Britain. Yet as a scientist who was a woman in post-war Oxford she had to struggle. She waited until 1960 to be appointed professor, though four years later she was awarded the Nobel Prize for Chemistry and became the first British woman to win a science Nobel.

Above: Dorothy Crowfoot Hodgkin, holding her penicillin model *c.*1940s.

Opposite: Molecular model of penicillin by Dorothy Crowfoot Hodgkin, England, 1945.

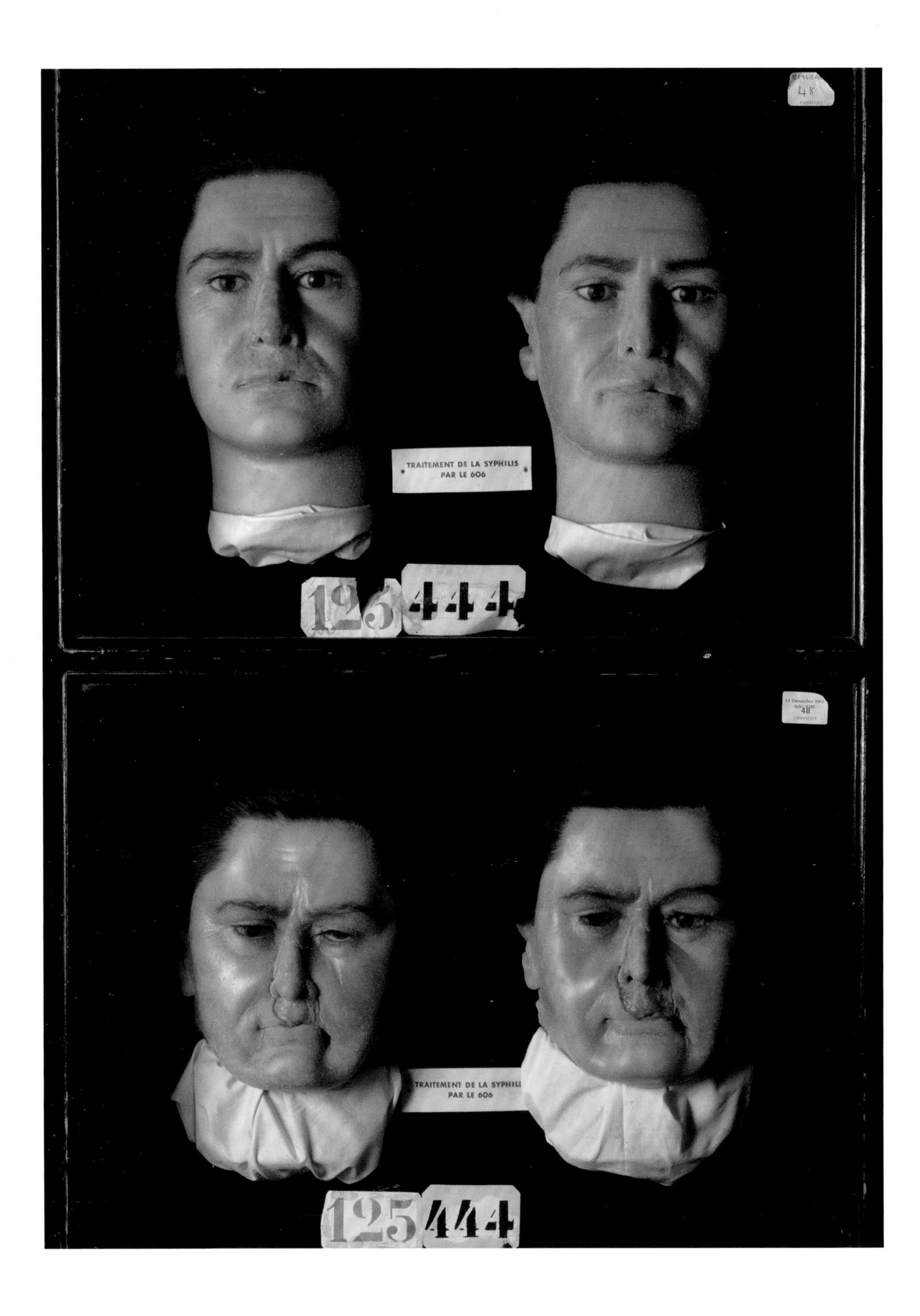
TRAITEMENT DE LA SYPHILIS
PAR LE 606
125 444
TRAITEMENT DE LA SYPHILI
PAR LE 606
125 444

WAX MODELS SHOWING A TREATMENT FOR SYPHILIS

Swiss artist Léonce Schiffmann used these heads demonstrating the effects of syphilis as part of his travelling display of anatomical waxworks.

The ways in which information about disease is communicated shapes public understanding and experience of health and illness. These disturbing wax heads were made to illustrate the successful use of the drug Salvarsan 606 as a treatment for syphilis, an infection transmitted by sexual intercourse or from a pregnant woman to her unborn child.

Salvarsan 606 was discovered as a treatment for syphilis by Paul Ehrlich and Sahachiro Hata between 1909 and 1912. Ehrlich called his drug a "magic bullet", referencing the drug's ability to target and kill specific disease-causing germs without harming the rest of the body. However, the drug did not take effect immediately and required multiple painful injections. Furthermore, it was an arsenic-based compound and therefore toxic to humans. Despite its shortcomings, the magic bullet was the most effective treatment ever developed for the treatment of syphilis, which had caused great suffering to countless men, women and children over a period of more than 500 years.

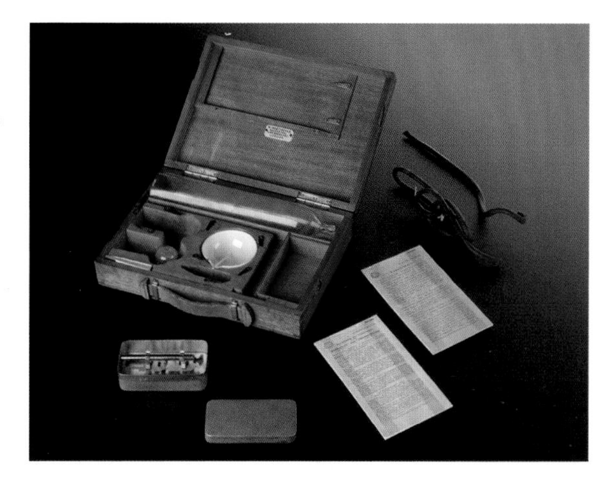

Opposite: Four wax heads showing the effects of syphilis and their treatment, 1910–20.

Above: Salvarsan kit for the treatment of syphilis, 1912.

Made in Germany in the early twentieth century, these morbidly fascinating wax heads were part of a tradition of creating anatomical waxes for public display, which stretched back hundreds of years. Such models were intended to educate the general population about anatomy and disease, but they were also designed to entertain – and, ultimately, to sell tickets. Shortly after they were produced, the heads were purchased by Léonce Schiffmann, an artist from Switzerland, for inclusion in his touring exhibition of anatomical waxwork models entitled *The Wonder of Life*. When Schiffmann died, his collection was passed on to Lily Binda and William Bonardo, the owners of a travelling public display, known at the time as a "freak show". Such shows, which displayed models, people and animals with medical conditions or bodies considered to be different, were outlawed in Europe in the 1960s.

The heads reflected and reinforced popular perceptions of syphilis and sexually transmitted diseases in the early twentieth century. At the time syphilis was heavily intertwined with ideas of social and moral degeneracy. The first three faces are scarred and contorted by pain and misery, destined to live as reviled outcasts. The final head – the happy ending, the patient cured – is clean and composed, ready to rejoin society. The theatrical exhibition of the syphilitic sufferer, at fairgrounds and travelling shows, served as a warning against the dangers of sexual activity.

"TABLOID" OPIUM POWDER

Collected from the seeds of the poppy, opium's powerful narcotic properties are highly addictive.

Extracted from the seeds of the beautiful poppy, opium was used as a remedy in Asia for centuries, only spreading to the West in the sixteenth century following the establishment of new trade links. Paregoric (liquid opium) appeared as a brand-new remedy in the 1788 book *Pharmacopoeia,* and this powerful narcotic quickly became enormously popular. By the 1820s the chemical version of opium, morphia (named for the Greek god of dreams), was available on the market, sold over the counter or as a popular prescription given by physicians to treat everything from insomnia to fever. Different types of opium were made and sold by Burroughs, Wellcome & Co, one of the first large pharmaceutical manufacturers in Britain. The bottle illustrated dates from 1938 and once contained aromatic chalk powder mixed with opium, a standby in all travelling medicine chests to treat dysentery and diarrhoea.

The longing to ease our pain is one of the most intense we can encounter. But when a painkiller has strongly addictive properties, a physical as well as emotional reliance gives the medication a huge control over us. Opium, the most famous of all narcotics, has been used for a variety of medicinal and recreational reasons and, like many addictive drugs, has a complex identity. A raddled drug addict helplessly smoking in an opium den was a common feature in Victorian melodrama, but a upper class lady dosing herself with her laudanum "drops" would have been the height of respectability. Laudanum was the liquid form of opium and morphine mixed with alcohol, highly addictive and yet as acceptable as sipping a pre-dinner glass of sherry. The use of opium rocketed throughout the Victorian age, but even by the late-eighteenth century opium's addictive qualities were suspected, although an addiction to drugs was seen as an indication of a weak personality rather than due to the incredibly addictive nature of opium. Since the mid-nineteenth century, laws to control access to opium have been passed in Britain with varying degrees of success. Since 1971 tincture of opium has been listed as a class A drug – deemed the most dangerous to use, and as such, its sale and use penalized most harshly.

Left: *Papaver somniferum* (Opium poppy).

Opposite: Empty bottle with glass stopper used for aromatic chalk powder with opium, 1937–38.

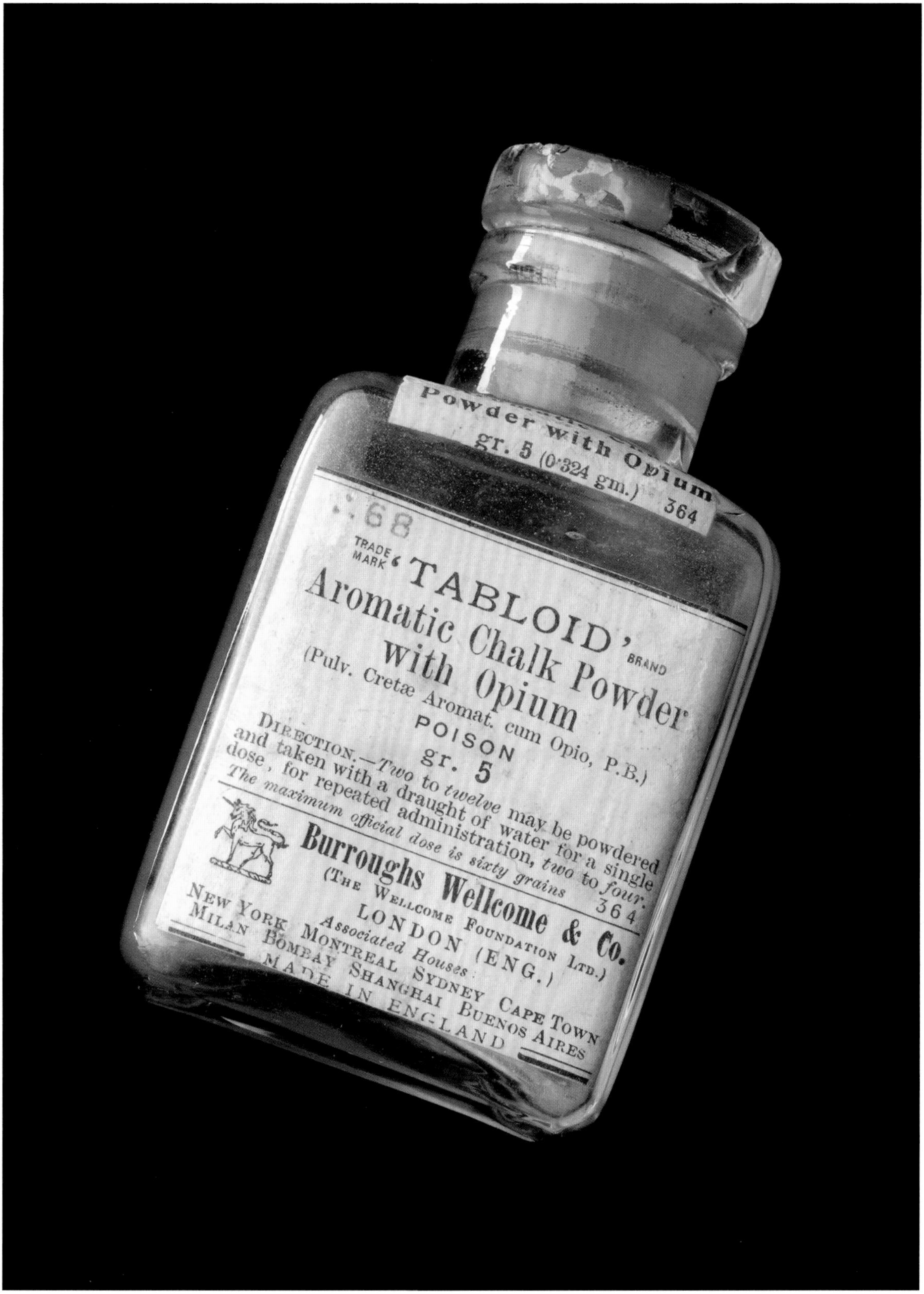

Powder with Opium
gr. 5 (0·324 gm.) 364
68
TRADE MARK 'TABLOID' BRAND
Aromatic Chalk Powder
with Opium
(Pulv. Cretæ Aromat. cum Opio, P.B.)
POISON
gr. 5
DIRECTION.—Two to twelve may be powdered and taken with a draught of water for a single dose, for repeated administration, two to four.
The maximum official dose is sixty grains 364
Burroughs Wellcome & Co.
(The Wellcome Foundation Ltd.)
LONDON (ENG.)
Associated Houses:
New York Montreal Sydney Cape Town
Milan Bombay Shanghai Buenos Aires
MADE IN ENGLAND

This Indenture witnesseth, That *Thomas Geo. Slaughter Son of John Slaughter of Northiam Sussex* [illegible] *Surgeon*

doth put himself Apprentice to *Richard Painter of Broadway Westminster Apothecary*

to learn his Art; and with him (after the manner of an Apprentice) to serve from the Date hereof until the full End and Term of Seven Years, from thence next following, to be fully complete and ended. During which Term, the said Apprentice his said Master faithfully shall serve, his Secrets keep, his lawful Commands every where gladly do. He shall do no Damage to his said Master, nor see it to be done of Others, but that he to his Power shall let or forthwith give Warning to his said Master of the same. He shall not waste the Goods of his said Master, nor lend them unlawfully to any. He shall not commit Fornication, nor contract Matrimony within the said Term. He shall not play at Cards, Dice, Tables, or any other unlawful Games, whereby his said Master may have any Loss. With his own Goods or others, during the said Term, without License of his said Master, he shall neither buy or sell. He shall not haunt Taverns or Play-houses, nor absent himself from his said Master's Service Day or Night unlawfully. But in all Things, as a faithful Apprentice he shall behave himself towards his said Master, and all his, during the said Term. And (the Sum of *One hundred and ninety five Pounds* of lawful Money of *Great Britain*, being paid or secured to the said Master, as the Consideration for taking his said Apprentice) the said Master his said Apprentice, in the same Arts which he useth, by the best Means that he can, shall teach and instruct or cause to be taught and instructed; finding to his said Apprentice, Meat, Drink, Apparel, Lodging, and all other Necessaries, according to the Custom of the City of *London*, during the said Term. And for the true Performance of all and every of the said Covenants and Agreements, either of the said Parties bindeth himself unto the other by these Presents. *In Witness* whereof, the Parties above-named to this Indenture have put their Hands and Seals, the *seventh* Day of *May* in the *third* Year of the Reign of our Sovereign Lord *George the fourth* by the Grace of God, of the United Kingdom of *Great Britain* and *Ireland*, King, Defender of the Faith, and in the Year of our Lord, One thousand Eight hundred and *twenty two*

Sealed and delivered (being first duly stamped) in the Presence of

Edmund Bacot

Thos. Geo. Slaughter

Richard Painter

AN APOTHECARY'S APPRENTICESHIP

Young Thomas Slaughter was apprenticed to Master apothecary Richard Painter, as this indenture document shows.

An indenture was a legally binding document committing a child (usually aged around 12) to a seven-year apprenticeship with a Master to learn a trade. Masters were fully qualified and usually very experienced practitioners, passing their wisdom to the next generation. The document pictured is an indenture for an apprenticeship in the art of apothecary, contracted on 3 May 1822 between Thomas George Slaughter, son of John Slaughter of Northam, Sussex, and Richard Painter of Broadway, Westminster. The document contains the rules that Thomas needed to observe while in the care of his Master; among these, it is stated that he was not allowed to "contract Matrimony, play at cards, dice, tables, or any other unlawful games". During his long and rigorous apprenticeship Thomas learned to recognize drugs and their use, make preparations, and dispense complicated prescriptions to patients and doctors. Documents in our collection also show that he took a training in anatomy to further his medical knowledge, and that he became fully qualified and went on to become a Master himself.

At the time of the indenture, Thomas's father John paid the onerous sum of £195, equivalent to one farm labourer working for 10 years, for his son's apprenticeship to cover "meat, drink, apparel and all other necessities". The price may have been too high for many people, but John was a surgeon himself. The family was probably wealthy and keen to maintain its status as a medical family. It was common at the time for families to specialize in the same trade across several generations, and Thomas clearly came from a long line of medical men.

Originally apothecaries belonged to the livery Company of Grocers, who themselves could be traced back to the Guild of Pepperers and Spicers. They sold everything from perfumes, spices and wines to herbs and medicinal drugs; the word "apothecary" comes from the Greek *apothēkē* meaning "storehouse". By the mid-sixteenth century apothecaries became increasingly associated with the preparation and trade of medicinal substances. The Worshipful Society of Apothecaries was created in 1617, thus breaking away from the Grocers Company. In 1704 the House of Lords ruled that apothecaries could both prescribe and dispense medicines, and as such are considered as the precursors of present-day GPs.

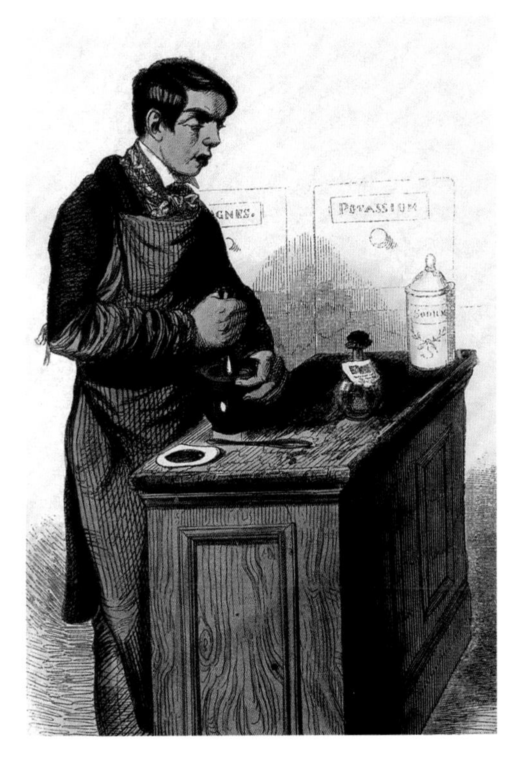

Above: A pharmacist's apprentice mixing up a prescription, nineteenth century.

Opposite: Apothecary indenture between Thomas George Slaughter and Richard Painter, England, 1822.

THE CASTLE OF PILLS

Created by the East London Health Project, this fairy-tale castle is constructed from colourful drugs.

The artists Peter Dunn and Loraine Leeson made this colourful castle from pills and drug packaging. It featured in a poster campaign commissioned by the East London Trades Council in 1978. Known as the East London Health Project, the initiative resulted in a series of posters that aimed to raise awareness about health-related issues among the local community, at a time when the National Health Service faced funding cuts. The posters were displayed in various medical venues, tackling subjects including mental health, the contraceptive pill and the pharmaceutical industry. They were also subsequently presented in art galleries, testifying to the conjunction of art and politics in these works.

The castle featured in a poster entitled *Passing the Buck, Games for Multinational Drugs Companies*, which was critical of the ways pharmaceutical companies operated. It shows a line of anonymous men dressed in suits and holding fists full of money, queuing on a coin-covered floor to enter a palace of pills. Three columns of text outline unethical profit-making tactics at the bottom of the poster (for example, "GREED BEFORE NEED"), while a dark and gloomy sky looms above.

This poster employs the use of photomontage, which involves the overlaying and recombining of images and text to create something new, often altering the meaning of the original images and presenting them in a new context. Photomontage has been used in the production of politically motivated and socially engaged work throughout the history of art, offering artists a means of creating powerful and critically loaded imagery. For instance, the German artist John Heartfield created a series of works using photomontage between the First and Second World Wars, reconfiguring text and images from the mass media to make anti-fascist statements and critique the political state of play. Leeson and Dunn have drawn from this heritage, creating socially engaged and politically orientated works in response to the health cuts proposed in the 1970s.

With the National Health Service still under threat today, the East London Health Project posters remain pertinent. As such, the posters have featured in recent exhibitions, including the Institute of Contemporary Art's *Things That Make You Sick* (2017), and in a BBC Four programme aired to commemorate the NHS at 70.

Left: *Passing the Buck, Games for Multinational Drugs Companies* (1970–80), East London Health Project.

Opposite: Exhibition model of a castle formed from various pills and drug containers, made by the East London Health Project, 1978–80.

DINCK · ANNO

PESTLE AND MORTAR

Every housewife would have owned a pestle and mortar specifically for grinding ingredients to create medicines for her family.

Today you might find a pestle and mortar in the kitchen for preparing dinner, but you would not typically use them to make your own medicines. However, before medication began to be mass-produced by pharmaceutical companies, the pestle and mortar were used by both apothecaries and women in the home to produce different remedies.

Created in 1590, this bronze, bell-shaped mortar and double-ended pestle may once have been used to crush and grind medicinal herbs and spices, or chemical ingredients such as mercury, into a powder or paste. Although pestles and mortars have been used for millennia within most cultures, this style of mortar was very common throughout Europe at this time, particularly the Netherlands.

These instruments were often used by apothecaries when preparing different remedies, from distilled waters to poultices. Apothecaries, who produced and sold medicines to patients and physicians alike, are often considered the predecessors of the general practitioners (GPs) we visit today. Passing down their knowledge through apprenticeships, their services were more affordable than those of university-educated physicians. As a result they gradually became the most widespread form of medical practitioner, forming the Society of Apothecaries in 1617. Used to prepare remedies for a range of afflictions and ailments, the pestle and mortar were important tools, and were often used as symbols to advertise the apothecary's wares.

The pestle and mortar were also tools used to create home-made remedies for a variety of ailments. The services of both physicians and apothecaries could be expensive, and many men and women preferred to create their own treatments. In particular it was the responsibility of housewives to preserve the welfare of their household, and they were often given pestles and mortars as wedding gifts. Many compiled collections of recipes that had been passed down through the family, shared by friends or copied from household manuals, almanacs and pharmacopoeias. Utilizing medicinal herbs grown in the garden or ingredients purchased from an apothecary, housewives used equipment found in their kitchens to create their own remedies for everyday diseases and injuries.

Whether used by the apothecary or in the kitchen, these ubiquitous objects clearly had an important role to play in a world where medical knowledge and managing ill health were deeply ingrained in everyday life.

Above: An eighteenth-century housewife's recipe for a "most excellent medicine against the plague". This recipe contains sage, rue, nutmeg and ginger, several of which would need to be prepared using a pestle and mortar.

Opposite: Mortar and pestle wedding gift, engraved "Love overcomes all things 1590".

PHARMACY JAR FOR LEECHES

Leeches used for bleeding a patient are not the prettiest of animals, but their ornate storage jars graced the pharmacy shelves.

Though beautiful, this jar was home to blood-thirsty leeches (*Hirudo medicinalis*) before they were sold to physicians and barber-surgeons. Once purchased, it was the leeches' job to feast on the blood of patients.

Based on the theory of the four humours, bloodletting was used for thousands of years to restore balance in the body. It was believed that having too much of one of the four humours (blood, yellow bile, black bile and phlegm) led to sickness. As blood was wet and warm, it was blamed for fevers and sweats. Bloodletting was performed with a variety of usually sharp tools, but the far-gentler leech has become something of an icon.

Bloodletting's popularity peaked in the 1800s, when demand for leeches was said to have outstripped supply. You may find yourself wondering how pharmacists maintained their stock levels of live leeches; it was only possible through the work of leech collectors who used their legs as bait, picking the leech off after it had fed on their blood. This takes around 20 minutes but, due to the enzyme in the leech's saliva, their bites continue to bleed for around 10 hours. Though this role came with some severe occupational hazards (namely blood-loss and infection), leech collectors were very poorly paid.

We may look back and consider this practice as a useless endeavour, but the idea of balance continues to dominate our discussions about health and medicine. We regularly talk about a balanced diet and lifestyle, but we just go about it in a slightly different way, and there are some very specific medical scenarios where we continue to use leeches.

Though the contents of the jar may invoke feelings of disgust in some people, the outside is undeniably beautiful. When not being handled by the pharmacist, this jar was part of the pharmacy's decorative display of products. Made some time between 1831 and 1859 by Samuel Alcock and Company, it is clear from the ornate design that this object was intended to impress and belonged in pride of place on the pharmacy shelf. But considering the appearance of leeches, is it any surprise that pharmacists chose to hide them from their customers?

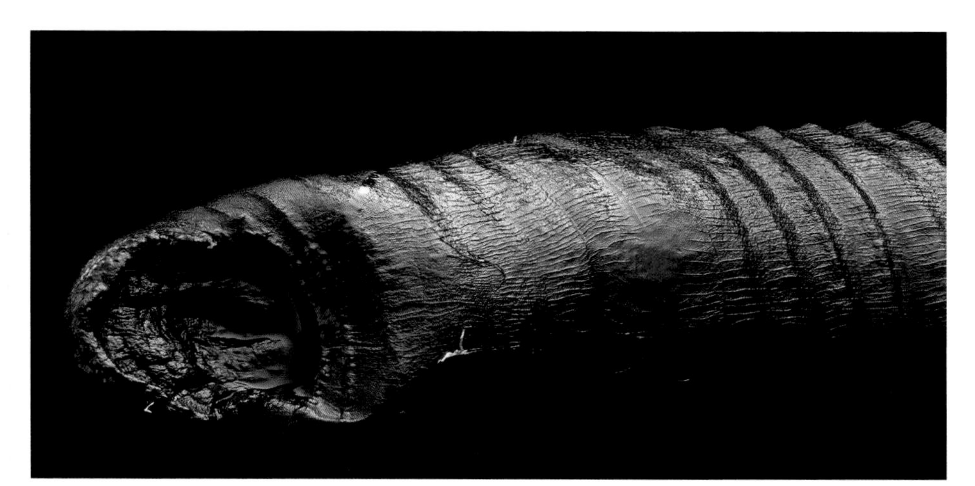

Oppposite: Ornate pharmacy leech jar, 1831–59.

Left: The mouth of a medicinal leech.

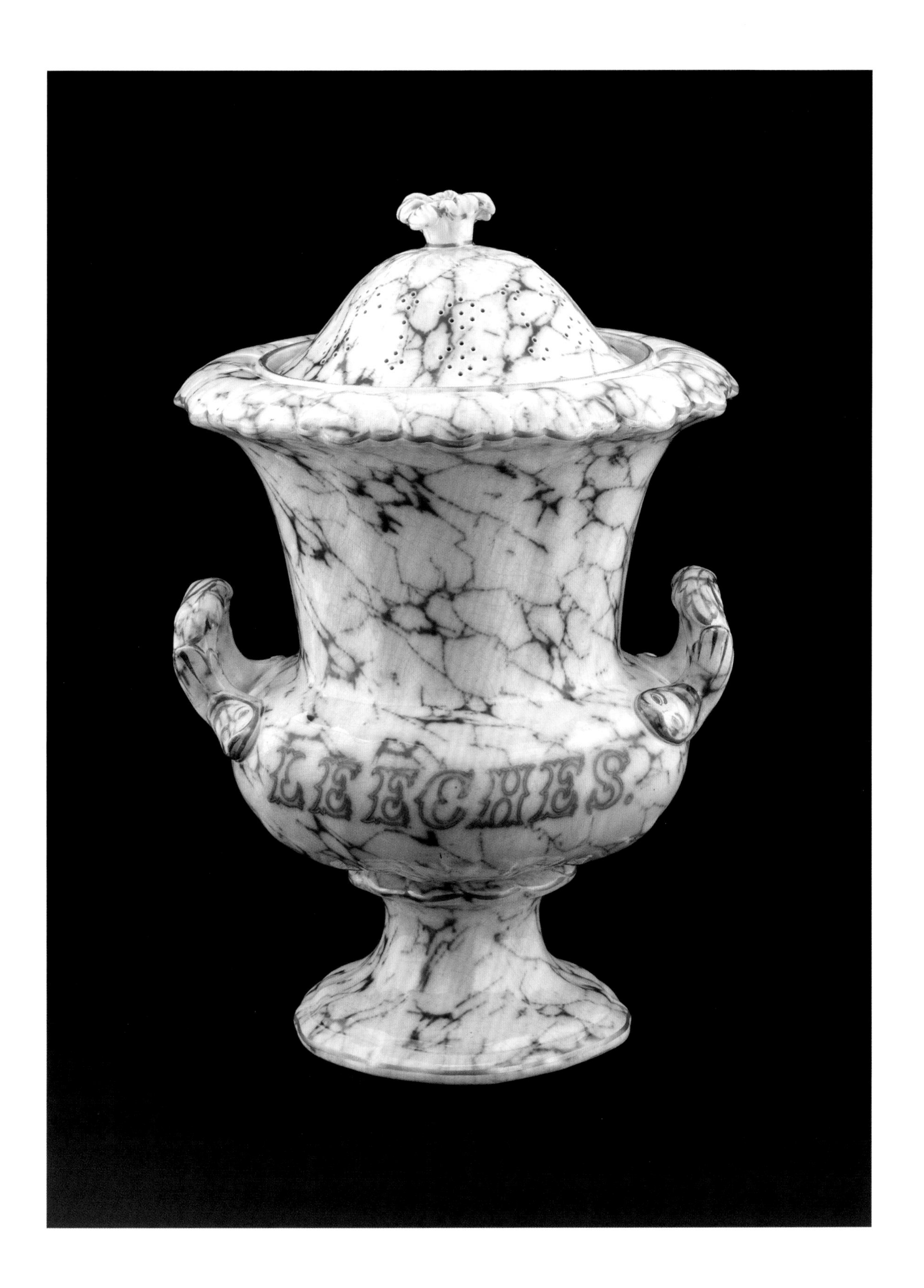
LEECHES.

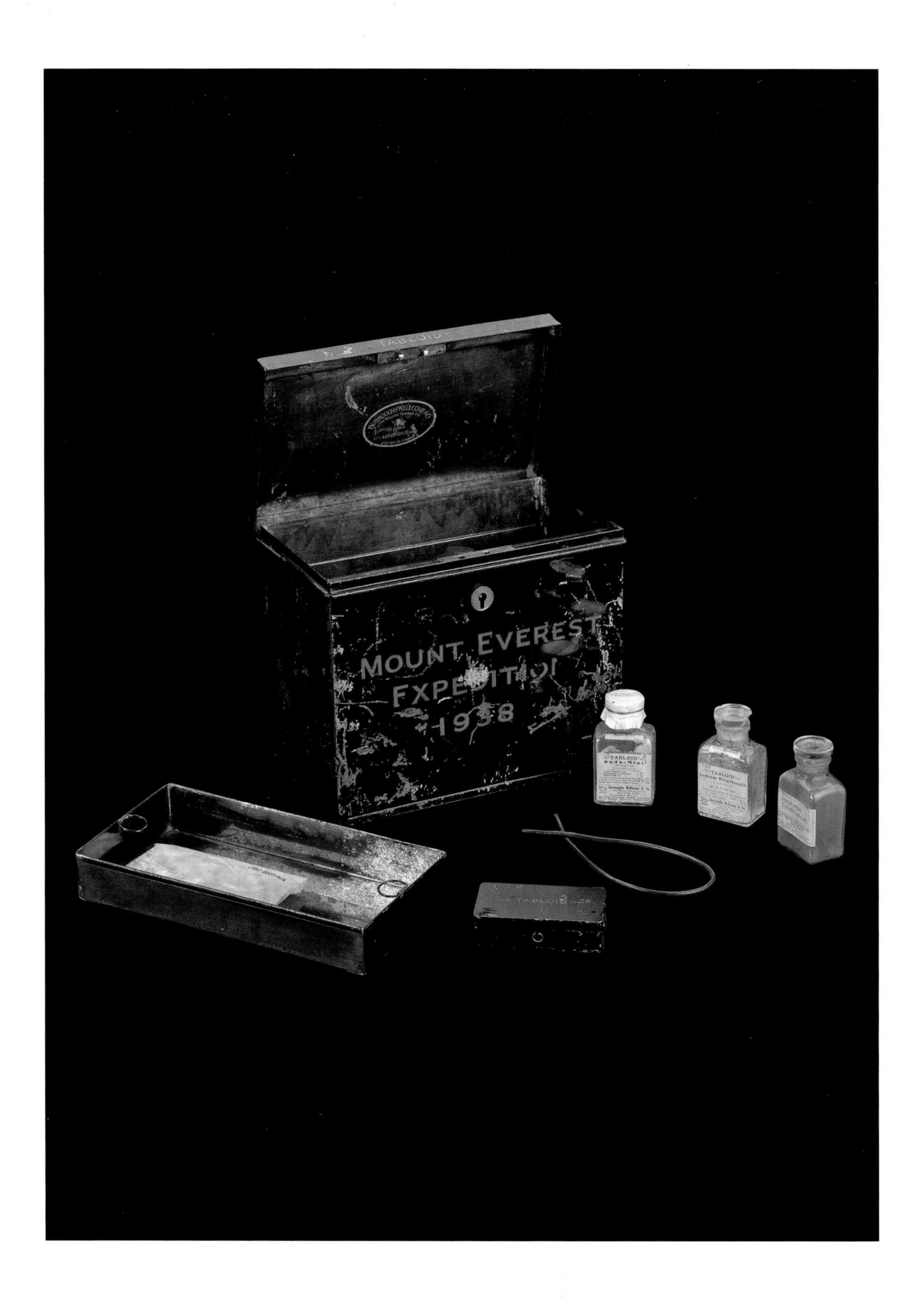
TABLOID
MOUNT EVEREST
1938
TABLOID

FIRST AID KIT USED ON MOUNT EVEREST

Henry Wellcome made sure his famous "Tabloid" medicine chests were given to every high-profile expedition for the advertising benefits.

Emblazoned with the words "Mount Everest Expedition 1938", this tin has taken many knocks on its 27,000-ft (8,230-metre) journey up one of the world's most famous mountains. This aluminium kit was used by the British Land Expedition when attempting to scale Mount Everest in 1938. Believed at the time to be the tallest mountain in the world at 29,029 ft (8,848 metres), a succession of British Expedition teams had tried and failed to reach the summit. High altitude and low oxygen levels proved extremely dangerous, placing an enormous toll on the human body, particularly when these expeditions reached the "Death Zone" at 26,000 ft (7,925 metres).

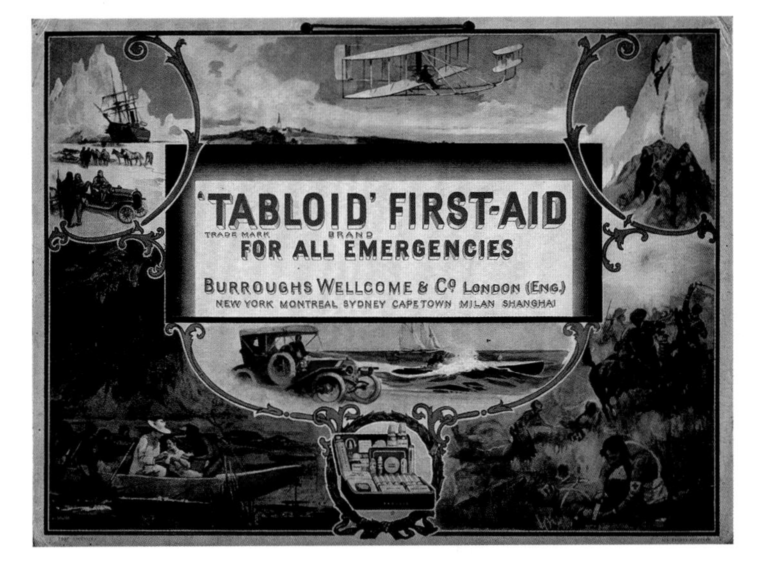

Opposite: Mount Everest expedition first aid case, 1938.

Above: Advertisement for "Tabloid" first aid kits, 1909.

Filled with medication to treat blood-poisoning, seizures, diarrhoea and respiratory problems – all of which are common ailments when reaching high altitudes – this chest was designed to keep the team alive and well for as long as possible during this 1938 mission. Unfortunately it was not to be. The expedition only reached 27,000 ft (8,230 metres) due to bad weather and sickness. However, the desire to reach the summit survived, and in 1953 these attempts finally met with success. Perhaps, therefore, this chest embodies the continued determination to reach what was then thought of as the top of the world.

However, it is also one of the many medical travel kits made by the pharmaceutical company Burroughs, Wellcome & Co and is stocked with bottles of the company's signature "Tabloid" pills. These pills, ready-made and easy to squeeze into a relatively small space, meant that a greater array of medical supplies could be carried, preparing the owner "for all emergencies". Henry Wellcome, who became the sole owner of the company in 1895, was central to both producing and marketing these wares. Ever the savvy businessman, he equipped many of the illustrious explorers of the day with medicine chests similar to this one, and exploited their fame to market the company's other first aid kits to more mainstream travellers. These battered chests would be returned and displayed at trade fairs and exhibitions, demonstrating the robust, convenient and high-quality nature of the company's products. For this reason many of these chests have survived, documenting the growing desire to reach the most extreme corners of the earth.

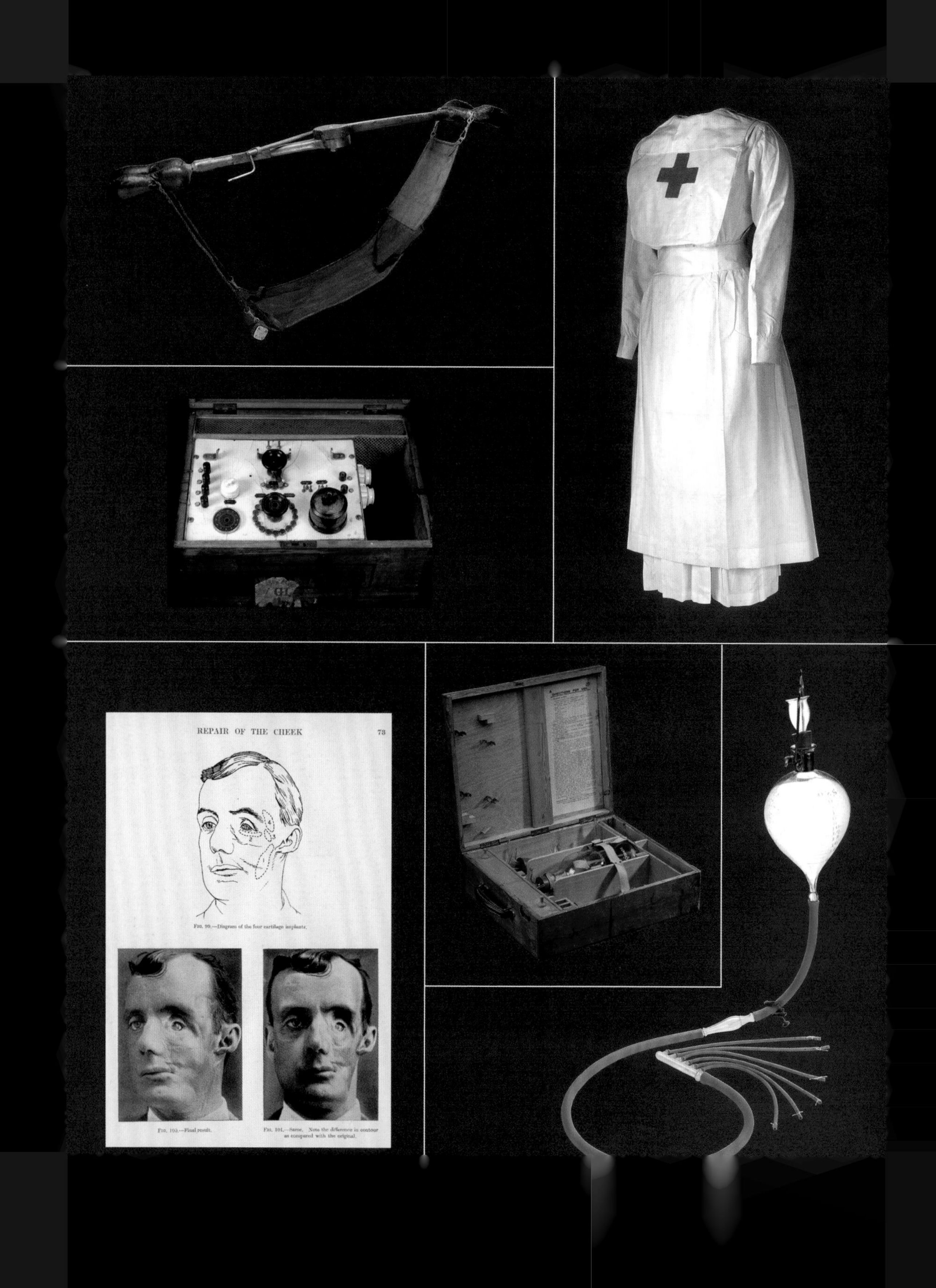
REPAIR OF THE CHEEK 73

Fig. 99.—Diagram of the four cartilage implants.

Fig. 100.—Final result.

Fig. 101.—Same. Note the difference in contour as compared with the original.

9

WAR

Throughout history, military conflict has produced an incalculable amount of human misery. From infectious diseases, spread like wildfire by mobilized troops, to the damage inflicted on bodies by deadly modern weaponry, war has had a direct and terrible impact on our health. Soldiers and civilians alike suffer from food deprivation and from the long-lasting effects of trauma. But the theatre of war has also provided enormous opportunities for medical development and breakthroughs in our understanding of health. Lack of supplies and time meant that medical practitioners can be forced to experiment and try new techniques, while the sheer number of sick and wounded humans meant outcomes of new treatments were easy to measure. Perhaps, above all, war can provide us with the strongest illustrations of human ingenuity and our determination to alleviate suffering.

FIRST WORLD WAR NURSE'S UNIFORM

Volunteer nurses played a crucial role in caring for the wounded on the battlefields of the First World War.

The crisp white folds of a nurse's cap and apron are more than a just protective layer against infection; they are a key part of the historic identity of a nurse, and perhaps never more so than with the nursing uniforms of the First World War. Shortly after the outbreak of hostilities in August 1914, a campaign was launched to recruit large numbers of volunteer nurses. It was recognized that the current professional or reserved occupation nurses were not enough for the forthcoming military action. For middle- and upper-class young British women, joining the Voluntary Aid Detachment (VAD) was often their first taste of freedom and opportunity to travel. VADs came from privileged backgrounds, compared to the pre-war professional nurses who had to work for their living. To differentiate them from professional nurses, VADs wore a large square white head scarf, folded and tied low over the forehead, which was to become one of the iconic images of the First World War.

The connection between uniform and identity for a nurse was recognized by the most famous nurse in history, Florence Nightingale. She insisted that her nurses wore identifying sashes and caps in the wards of Scutari Barracks during the Crimean War (1853–56), in effect labelling them as respectable, working women. This was especially important as, before Nightingale's nursing reforms, Victorian nurses had a poor reputation, often being seen as dirty and dishonest. Nursing work had been a free-flowing activity, carried out in the home or by travelling women. Most practical nursing within a hospital environment was carried out by nuns in nursing orders. The nurse's white apron and cap thus provided a physical association with the historic connections to religious orders and domestic servants.

This good-quality cotton uniform for a VAD, in the Science Museum's collection, was made by the upmarket department store Harrods, and has a beautifully embroidered red cross on the bib and head scarf. Its owner served at the Hôpital Anglais in Nice, France, during the First World War. In this period both volunteer and professional nurses took on extra duties, becoming unofficial surgical assistants and anaesthetists, in addition to more general military nursing work. The apron's now-pristine condition belies the messy and challenging work that its former owner carried out, providing care and comfort to her military patients.

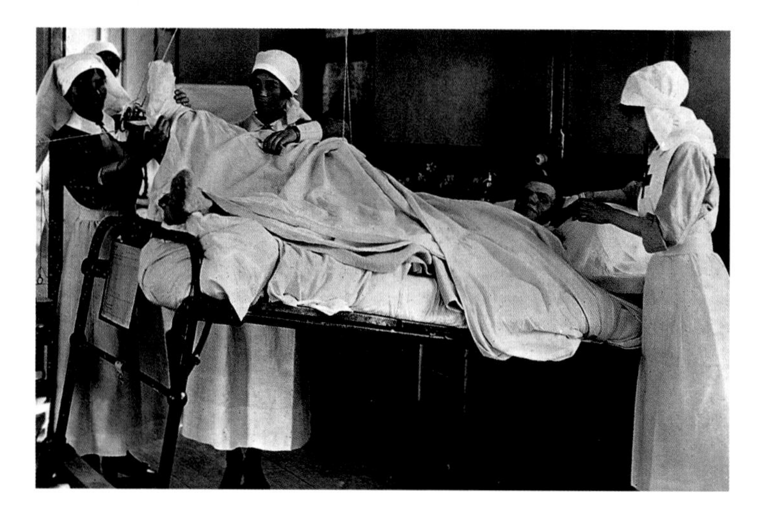

Left: Three nurses dress the wounds of a patient on a hospital ward, 1914–18.

Opposite: One round-bib, white cotton apron with cross straps, and with fabric red cross sewn into bib, worn in France, 1914–18.

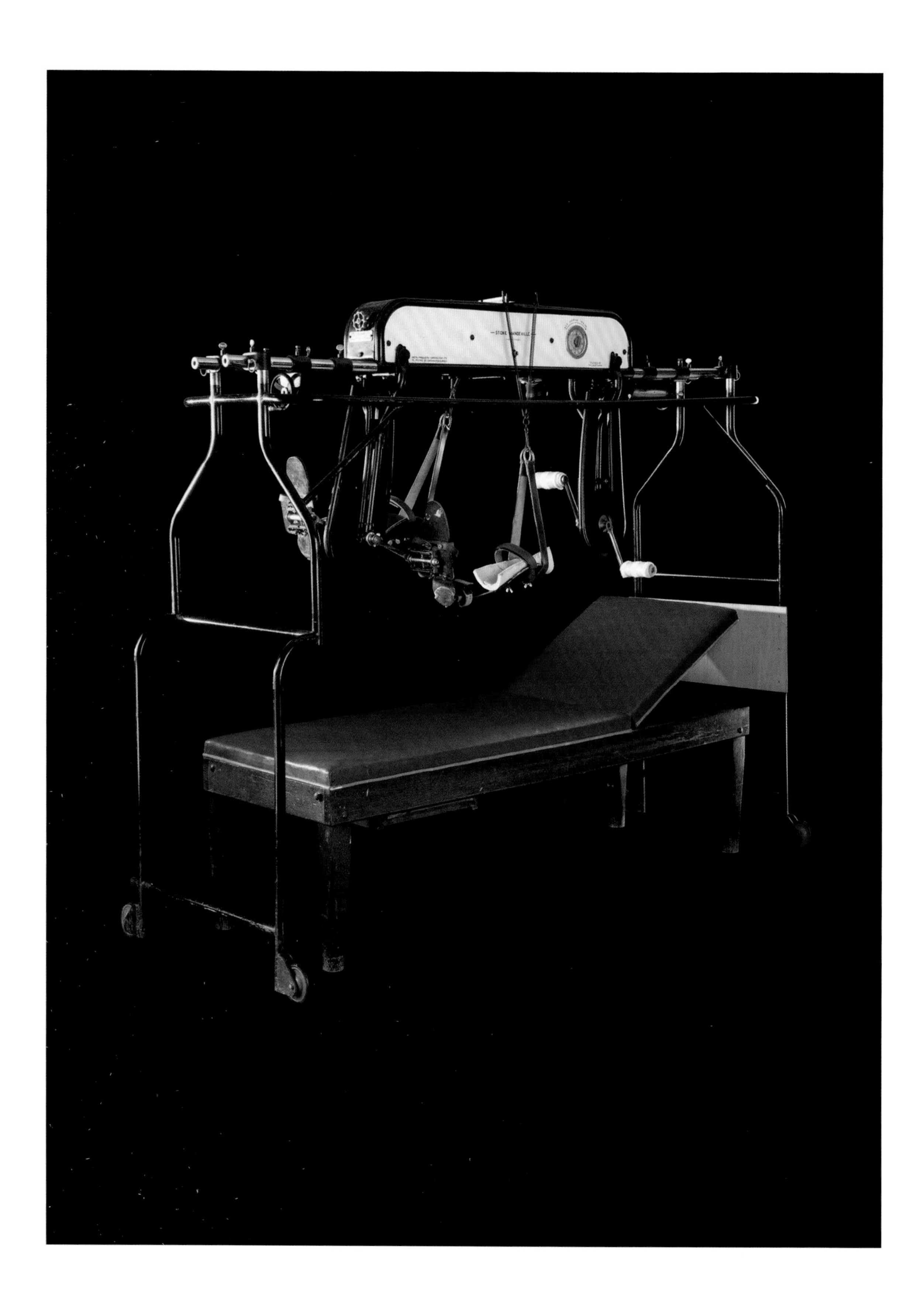
— STOKE MANDEVILLE —

THE BED CYCLE

Ludwig Guttman recognized the importance of exercise for his patients with paralysis, and this work led to the founding of the Paralympic Games.

Keep pedalling, keep going, only three minutes to go! Known as a bed cycle, this contraption was designed by German neurologist Ludwig Guttman during his time at Stoke Mandeville Hospital in Buckinghamshire, England. From March 1944, Guttmann ran the newly established specialist spinal unit for wounded soldiers returning from the Second World War. Using techniques from Europe and the United States, he set about changing how people with paralysis could be treated and to improve the length and quality of their lives.

Guttman viewed exercise and physiotherapy as essential to build physical strength. With options to change the tension of the springs, a timer of up to 60 minutes and a speedometer in miles per hour to see how fast they could go and encouraged to use this machine to exercise their arms and legs. The metal frame with makeshift pedals could be wheeled over a bed, and users adapted the bike by taping bandages to the hand pedals for comfort.

Exercise was so important to daily life at Stoke Mandeville Hospital that, on the same day that the 1948 Olympic Games began in London, 16 ex-service personnel competed in the first ever Stoke Mandeville Games, in just one sport – archery. More sports were added every year and, in 1952, the first international competitors arrived. The ninth International Stoke Mandeville Games, held in Rome in 1960, are now considered to be the first Paralympic Games. The word "paralympic" comes from Guttman's vision to see his Games as a parallel to the Olympic Games. From his retirement in 1966 until his death, Guttmann remained heavily involved with the Paralympics.

This particular piece of apparatus was made for and used at The Chaseley Trust in Eastbourne, Sussex. Originally called The Chaseley Trust for Disabled Ex-Servicemen, it was set up as a residential home at the request of Ludwig Guttman. The bed cycle was in use from 1949 and was last used in 1994 by a person with tetraplegia – the loss of function in all limbs and torso. Five years later the Trust donated it to the Science Museum and, as far as anyone knows, it is the only surviving example. Looking at it, you can imagine the sweat, frustration and effort exerted on it by its users on their roads to recovery.

Opposite: Hospital-bed cycle made in 1949.

Above: Close-up of the bed cycle's timer.

WOUND IRRIGATION TREATMENT

Dripping sodium hypochlorite onto wounds saved countless lives in the First World War and changed the way injuries are cared for.

The extraordinary circumstances of military conflict can lead to dramatic medical breakthroughs. The sheer numbers needing treatment, the urgency and sometimes the financial investment in medical research can all lead to new discoveries being made at these times of human misery. The Carrel-Dakin method of wound irrigation is one such innovation, which changed the way that wounds were cared for in the early twentieth century.

The deadly modern weaponry on the Western Front of the First World War caused dramatic wounds. Infection was driven in through shrapnel, dirt, clothing or even bacteria from the rich fields of northern France and Belgium. Wounds were managed in the same way that they had been for centuries – cleaning, then pulling the two edges together, stitching ... and hoping for the best. This often resulted in the fatal "gas gangrene", when bubbles of gas in the tissue form as a reaction to bacteria in a wound. A tell-tale crackling noise is heard when pressure is applied, a noise that signified at best an amputation, at worst the death sentence of the unfortunate soldier.

British chemist Henry Dakin and French physician Alexis Carrel worked together to completely rethink the treatment of wounds. For the first time they addressed the concept of allowing a wound to heal from the bottom up. While the healing process took place, keeping the wound hydrated and, above all, clean was of primary importance. Henry Dakin was responsible for creating the hydrating fluid, a mixture of sodium hypochlorite, while Dr Carrel developed the method at his hospital, a former hotel, in Compiègne, northern France. Dr Carrel was a complex man: a recipient of the Nobel prize in 1912 for his remarkable work on heart surgery, although his enthusiasm for eugenics makes uneasy reading today.

The Carrel-Dakin method was simple but needed constant monitoring. Dakin's fluid was stored in a glass bottle above the patient's head, with rubber tubing connected to a glass tube with a number of rubber nozzles, which directed fluid onto the wound itself. Careful nursing was important in this treatment's success – cleaning, changing the tubes and testing swabs for infection were all needed. Wound irrigation was one of the most important milestones in managing infection before the discovery of penicillin. Its use caused a dramatic reduction in the loss of life during the second half of the First World War.

Opposite: Carrel–Dakin irrigation performed by Millicent Sutherland-Leveson-Gower, the Duchess of Sutherland, and Dr Oswald Gayer Morgan at No. 9 Red Cross Hospital (Millicent Sutherland Ambulance) at Calais, 27 July 1917.

Below: Carrel's apparatus for sterilizing wounds with Dakin solution, used by the British Army in the First World War. The original rubber tubing has perished, and modern tubing is shown here.

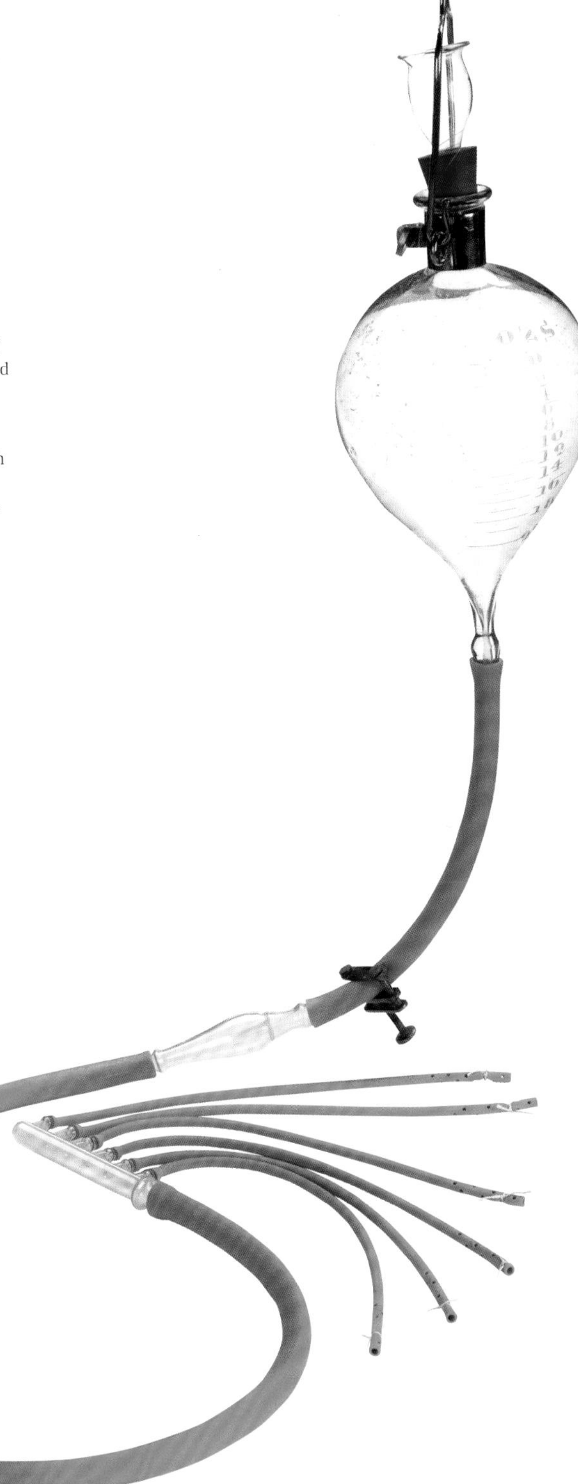

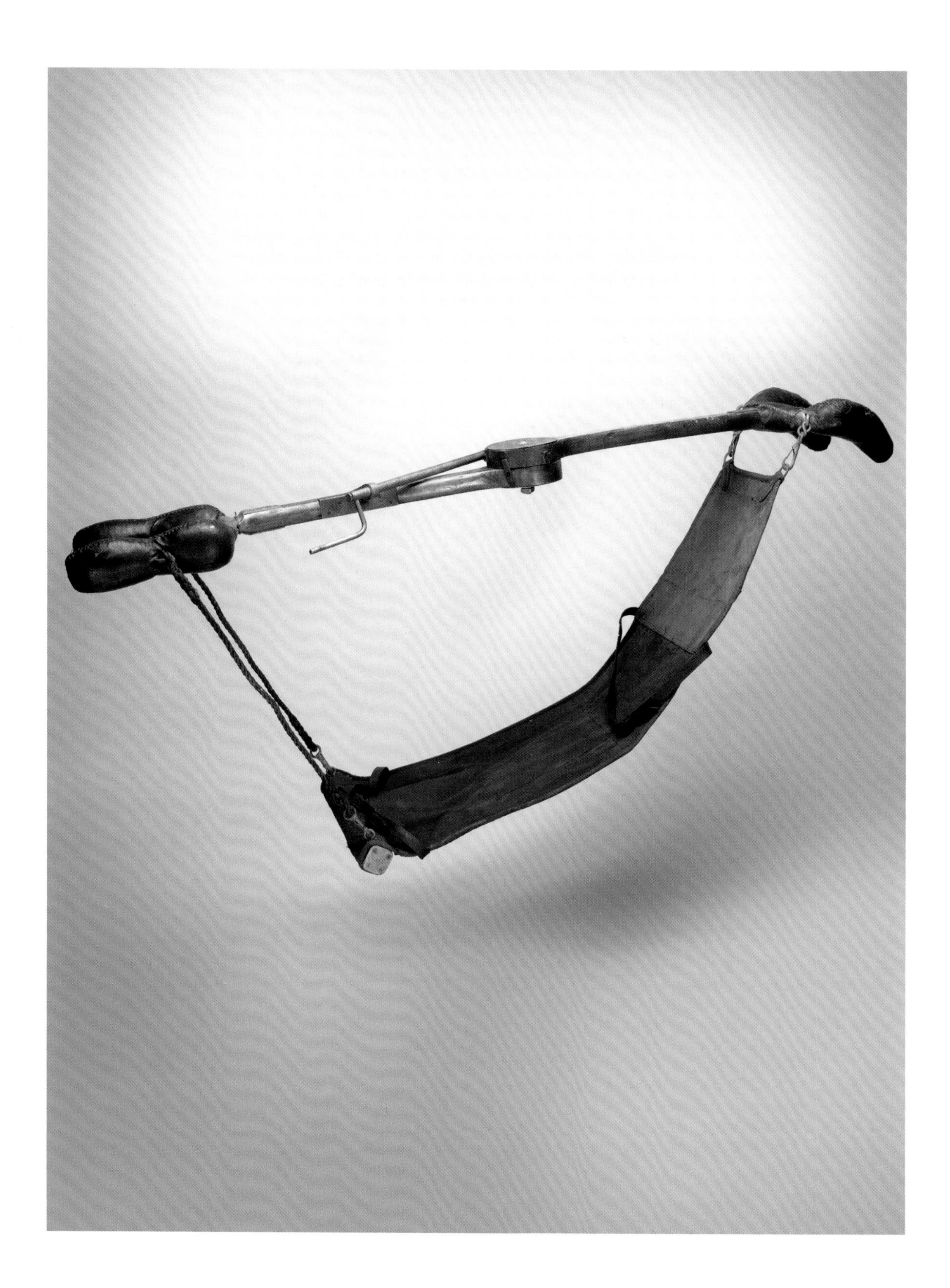

THE COLT STRETCHER

Surgeon George Colt's invention carried the wounded through the zig-zag layout of trenches in the First World War.

In certain contexts there could be something vaguely comical about the idea of a stretcher that can bend around corners. But the logic behind what might seem an eccentric and rather trivial proposal was deadly serious, and it was just one of a series of attempts made during the First World War to deal with a very specific problem. This was a new kind of war, which produced casualty levels unparalleled in scale and severity. Huge medical challenges were created, and these included the act of moving the wounded away from battlefields to places of greater safety, where they could receive treatment for their injuries.

For the British forces, a war that began as one of movement soon settled into a static war of attrition as combatants dug complex systems of opposing trenches, from which they launched occasional attacks. One characteristic of those trench systems nearest to the frontlines was that they zigzagged back and forth, rather than being dug in a straight line. This was a defensive technique to reduce the blast effect of a shell and the dangers of intruding enemy soldiers shooting straight down a long stretch.

One result of this was that movement through such trenches, which could also be extremely narrow, was both slow and arduous. This was especially so for stretcher bearers carrying a wounded man on a conventional stretcher. These were heavy, inflexible and had two rigid wooden poles that were several feet long. To make things harder, in the chaos of battle the movement of wounded men was not the top priority – getting supplies and reinforcements to where they were needed came first.

The British surgeon George Colt proposed one solution with this bendable design, known as the Colt stretcher. Carried on the bearers' shoulders, the wounded man would lie semi-upright and facing backwards in the canvas sling suspended underneath. Above him was a thick single wooden bar with a central pivot, the movement of which could be controlled by a handle turned by the bearer at the rear. How successful Colt's invention was is unclear, but this example has evidently been used and also modified, since extra padding has been added for the two bearers. But, successful or not, it is a poignant reminder of how technological ingenuity can often be driven by a humane desire to alleviate suffering in the most inhumane of environments.

Opposite: A version of Colt's stretcher for narrow trenches, 1915–18.

Above: The difficulties of moving a standard stretcher through narrow, winding trenches is clear in this image taken on the Western Front during the First World War.

FIRST WORLD WAR X-RAYS

X-rays were essential in the location of shrapnel in the bodies of wounded men on the First World War battlefields.

On the battlefields of the First World War (1914–18), the bodies of soldiers were extremely vulnerable. Artillery was the deadliest weapon in this conflict and became increasingly powerful over the course of the war. Those not killed could be left with complex wounds well beyond the capabilities of doctors in the field. Shrapnel, shell fragments and other objects could also be blasted deep into flesh, shattering bones but also taking in the infectious mud and debris of the frontline environment. An individual's survival very often depended on the means to locate and remove these objects, which had become hidden from view.

Given the huge levels of casualties, it is unsurprising that the First World War saw innovations in medicine as well as weaponry. X-ray imaging became a breakthrough technology, but it wasn't new. Discovered in 1895, X-ray technology had featured in previous military campaigns, but during the First World War it was used widely on both sides and, crucially, became increasingly portable. Once the Western Front battlefields became locked into static trench warfare, suites of X-ray machinery could be established in field hospitals, while adapted vehicles took the technology even further forward. X-rays had a range of uses, including the visualizing of fractures, but they were also crucial in both locating and removing the projectiles that lay deep inside a soldier's wounds.

Made in Austria, this mobile X-ray unit was used by medical teams based some distance behind German-held lines. The set contains the various component parts of an X-ray machine, as well as the plates and accessories needed to see the results. Packed into a series of wooden carrying cases, the unit could be quickly disassembled by a specialist group, transported, then reassembled in a new location if required. Such equipment played a hugely important medical role during the conflict, but regrettably it came at a cost. Over the course of the war, many medical personnel were routinely exposed to what we would now consider extremely hazardous levels of radiation.

Left: One of the series of equipment-filled boxes from a German field X-ray set.

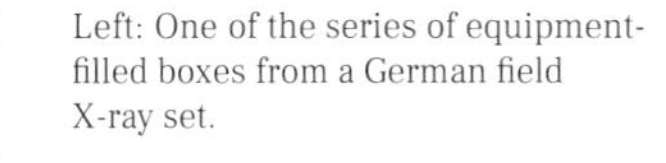

Opposite: X-ray taken at a British military hospital, showing a bullet embedded in the skull of Private A. Neal, September 1916.

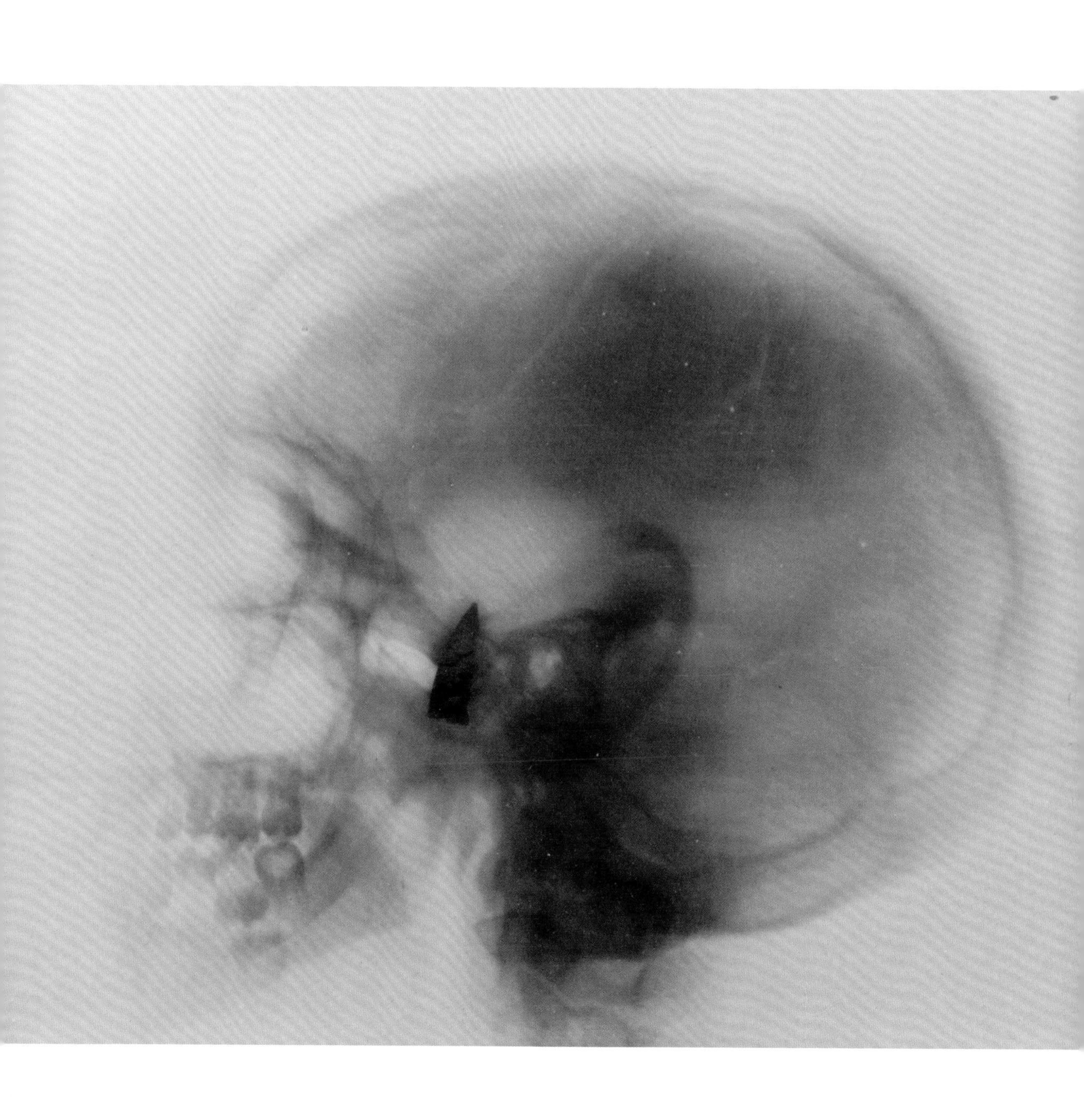

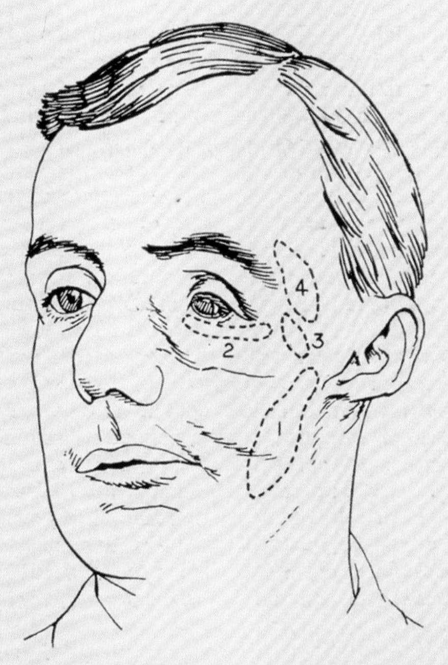

Fig. 99.—Diagram of the four cartilage implants.

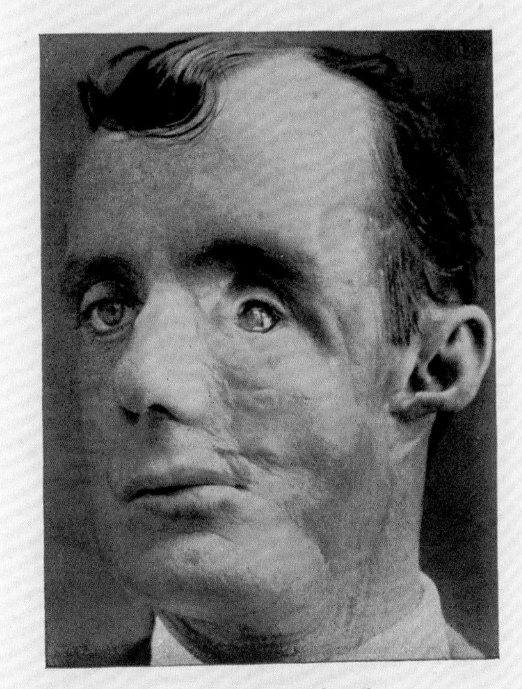

Fig. 100.—Final result.

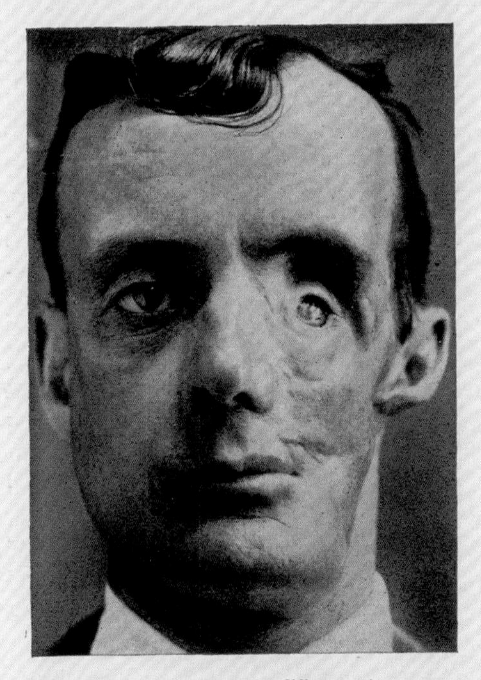

Fig. 101.—Same. Note the difference in contour as compared with the original.

GILLIES' *PLASTIC SURGERY OF THE FACE*

Harold Gillies pioneered facial plastic surgery, repairing the damaged faces of soldiers wounded in action.

Harold Gillies, born in Dunedin, New Zealand in 1882, moved to Britain to study medicine at Cambridge University and later became a surgeon. When the First World War broke out, he was sent to France, where he met Auguste Charles Valadier, a French-American dentist, who was replacing soldiers' jaws that had been lost in combat. Here Gillies also met American dental surgeon Bob Roberts, who taught him about jaw and mouth surgery; Gillies was hooked.

As the war raged, facial wounds continued to become more and more common. Soldiers' bodies were shielded from enemy fire by the parapets of the trenches, but once they peered over the top, they were targets for enemy snipers. The danger remained beyond the trenches as many men received facial injuries when artillery shells and shrapnel exploded nearby. Gillies, spurred on by the number of facial wounds, implored the authorities to create a plastic surgery unit.

By 1915 Gillies ran his own plastics unit; in 1916 he had a hospital in Aldershot; and by 1917 the Queen's Hospital opened in Sidcup. It was the world's first plastic surgery hospital and had 1,000 beds. It was here that Gillies completed his most influential work, drawing upon his impressive surgical and artistic skills to recreate the damaged faces of his patients, and relying on a large number of staff, including artists, to document patients' roads to recovery.

Before the First World War, the way a patient looked was always of secondary importance; however, Gillies aimed to restore the function and aesthetics of the facially wounded. This often required numerous operations, something which was very dangerous before the discovery of antibiotics. It was only through experience that Gillies learnt that patients required significant rest periods between operations, so many of his charges remained in his care for years.

Gillies turned his war experience into this book, *Plastic Surgery of the Face*, published in 1920, which laid the foundations for modern plastic surgery. Within it, Gillies and his team diligently recorded patient progress through photography. This book not only represents a landmark in medical practice and documentation, but it is a testament to Gillies's skill, intelligence and devotion to his craft.

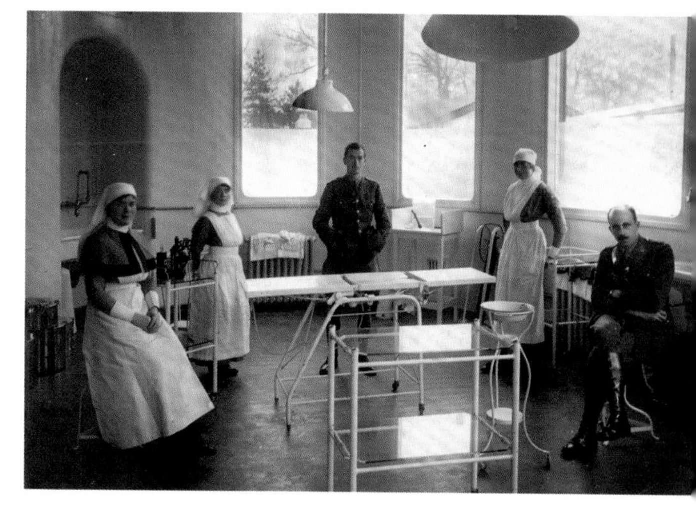

Opposite: *Plastic Surgery of the Face* documents Gillies's impressive and influential work restoring the faces of the injured, published in 1920.

Above: The Plastic Theatre, Queen Mary's Hospital, 1917. Harold Gillies is seated on the right.

FIRST WORLD WAR BLOOD TRANSFUSION KIT

Oswald Hope Robertson designed portable blood transfusion apparatus, which led to the development of the modern blood bank.

This jumble of bottles, boxes and tubing may not seem like much, but it was a lifesaver in the First World War. Oswald Hope Robertson, a British-born doctor serving with the US Army Medical Corps, designed this portable blood transfusion set. It relied on a system of syringes and cannulas to transfuse blood from a donor to the recipient but, unlike previous devices, it did not require a simultaneous connection between the two individuals.

We are so used to blood banks and the National Blood Transfusion Service that the fact of a donor needing to be so close to a patient may be surprising to us. Although it became possible to store blood in the latter stages of the First World War, it was very difficult to keep the blood fresh when close to the fighting fronts. The bottles used in Robertson's sets were pre-dosed with sodium citrate, an anticoagulant that prevented clotting, which meant the donated blood was viable for transfusion for several hours.

Blood transfusions had been attempted for hundreds of years before Robertson invented his kit. The first written account of blood transfusion dates back to the 1660s, when Richard Lower performed a blood transfusion on a pair of dogs. A few years later Lower transfused a Cambridge University student with the blood of a sheep, and surprisingly, the patient survived. Regardless of this success story, most people who underwent blood transfusions in this period died. It was not uncommon for practitioners to use animal blood in these procedures, which was often incompatible with most human blood types. It was not until 1901 that the first three human blood groups were discovered by Karl Landsteiner.

This discovery allowed practitioners to understand why some blood transfusions failed, enabling the practice to be improved upon. Blood transfusion took on a new importance during military campaigns. Even today, blood loss remains the most common cause of death on the battlefields, and the First World War was no exception. It was Robertson's work, along with that of other blood transfusion pioneers, that led to the routine use of transfusion in emergency situations and the establishment of the modern blood bank. Without their tenacity, ingenuity and experience, blood loss would pose an even more significant threat to our health.

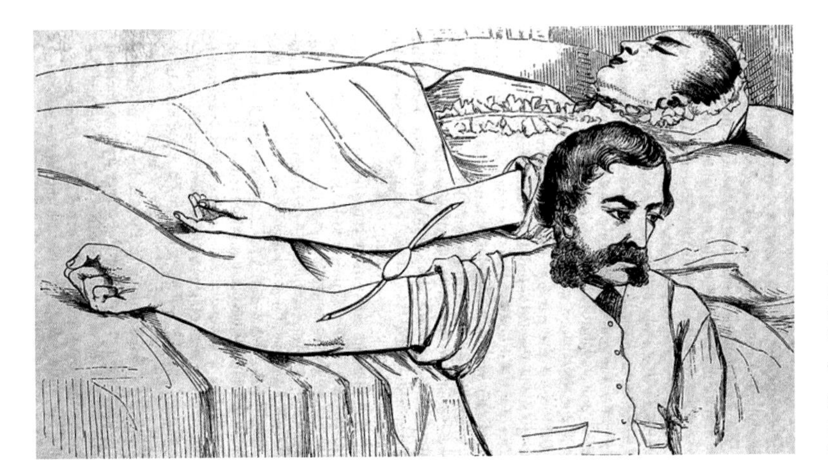

Left: Direct person-to-person blood transfusion for a woman during childbirth from J.H. Aveling's 'Immediate transfusion in England', *Obstetrics Journal*, 1873.

Opposite: This kit made blood transfusions possible near the fighting front, saving numerous lives during the First World War.

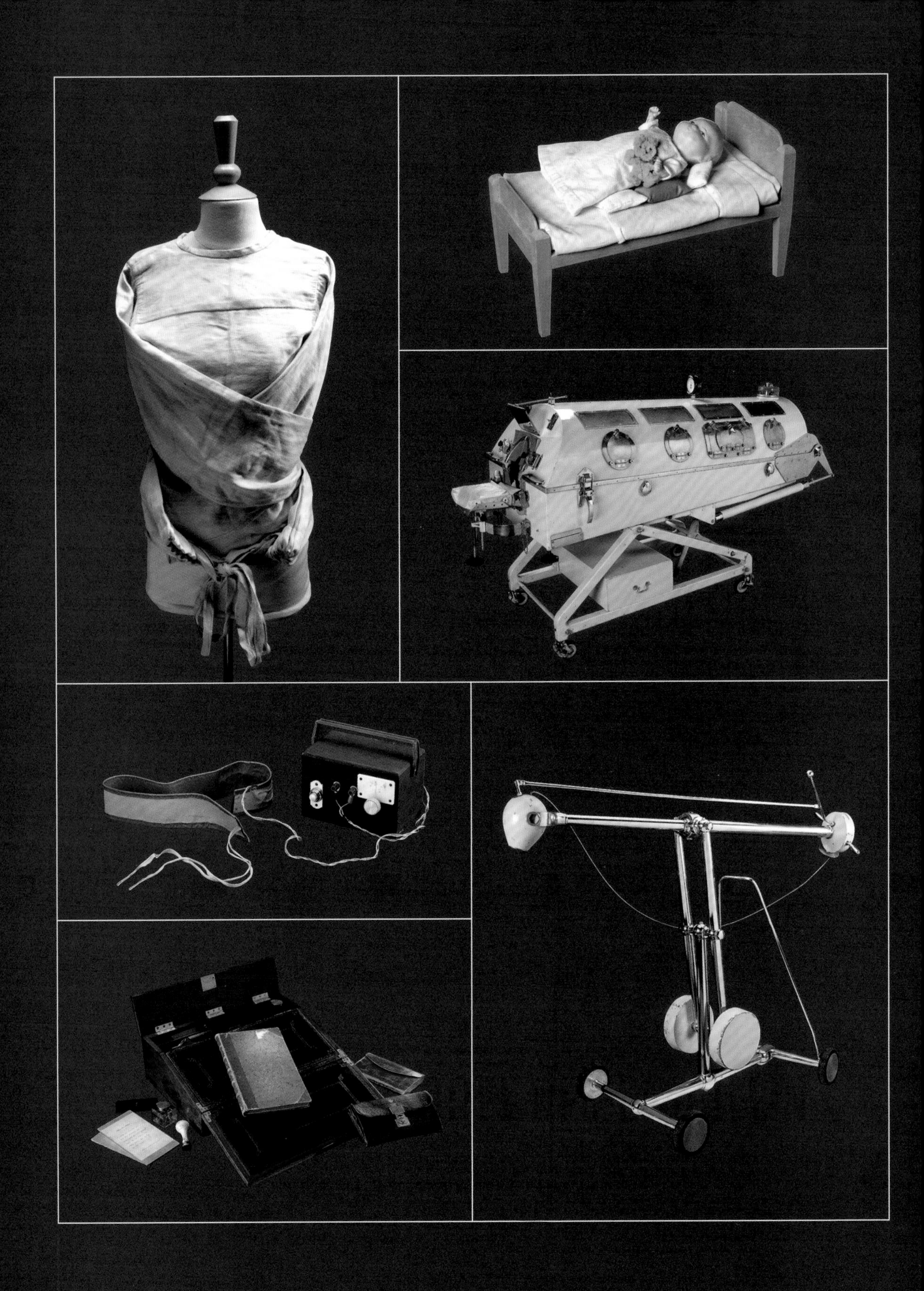

10

HOSPITALS

The modern hospital can trace its roots back to the religious infirmary attached to the local church for the poor of the parish, and echoes of the religious past can be found today in terminology such as calling nurses "Sister", a reminder of nursing nuns. Historically, hospitals were only a place for the very poor, used only by those who could not care for their loved ones at home. They could be dirty and dangerous places, and hospital-acquired infections are still a concern even today. Hospital furniture and apparatus can tell stories as effectively as the buildings themselves, from ordinary domestic furniture that might be found in any home to specialist, state-of-the-art equipment. Hospitals can create a physical record of conditions and treatments of the past, while aiming to become centres of medical research that will benefit future generations.

THE KING'S FUND BED

A collaboration with the Royal College of Art produced The King's Fund bed, easily adjusted to care for patients.

Walk into a general ward in any British hospital and you will be confronted by beds. They are the most familiar piece of hospital furniture, but they aren't like your bed at home – one of the most ubiquitous designs being The King's Fund bed. First designed in the 1960s, it revolutionized the role played by hospital beds and set new standards for hospital care. Previously, hundreds of different bed designs existed in British hospitals. The majority of them were little more than heavy, static "places of rest" with little scope for the movements and adjustments that could assist nursing staff with their work.

In a bid to standardize hospital beds and make them more user-friendly, a five-year project was run by The King's Fund charity, in collaboration with a design team led by Bruce Archer from the Royal College of Art's Industrial Design Research Unit. Archer took a controversial scientific approach, with extensive data collection, mathematical modelling and countless hours of observation of the ergonomics of hospital bed-related activity. Standardizing the hospital bed was seen as one way to improve nurse productivity and save money on procurement costs. In 1965, 20 prototype beds were installed for three months at Chase Farm Hospital in north London. Trained observers assessed their use from 6 a.m. to 10 p.m. each day, and their observations were collated to establish cost effectiveness, patient satisfaction and functionality for the frontline staff using them.

In 1967 The King's Fund published the new specification for the design of a hospital bedstead. The result was a bed that was easily moved, height-adjustable, had a reclining back-rest and could be easily tilted – changes activated by a series of foot pedals at floor level. The King's Fund beds were mass manufactured and purchased by hospitals around the country.

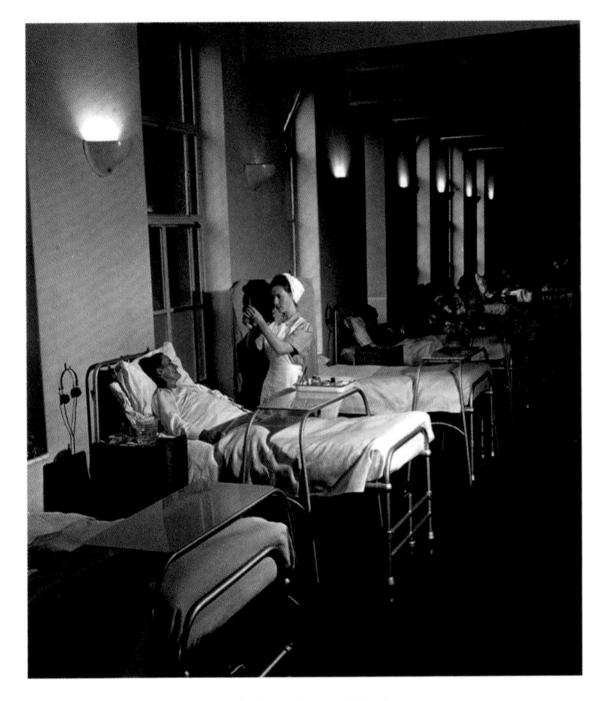

Above: Nurses tending to their patients in beds that were difficult to move, 1940s.

Opposite: A 1990s model of The King's Fund bed.

The King's Fund beds are still found in many hospitals, but over time the design has been tweaked and updated, and they are gradually making way for newer generations of more conveniently adjustable, electrically-powered, "push-button" beds. Hospital beds remain in the public eye with the constant concern about the flow of patients in and out of them, costs associated with them, and patient safety. The total number of National Health Service hospital beds in England has more than halved over the past 30 years, while the number of patients treated has increased significantly.

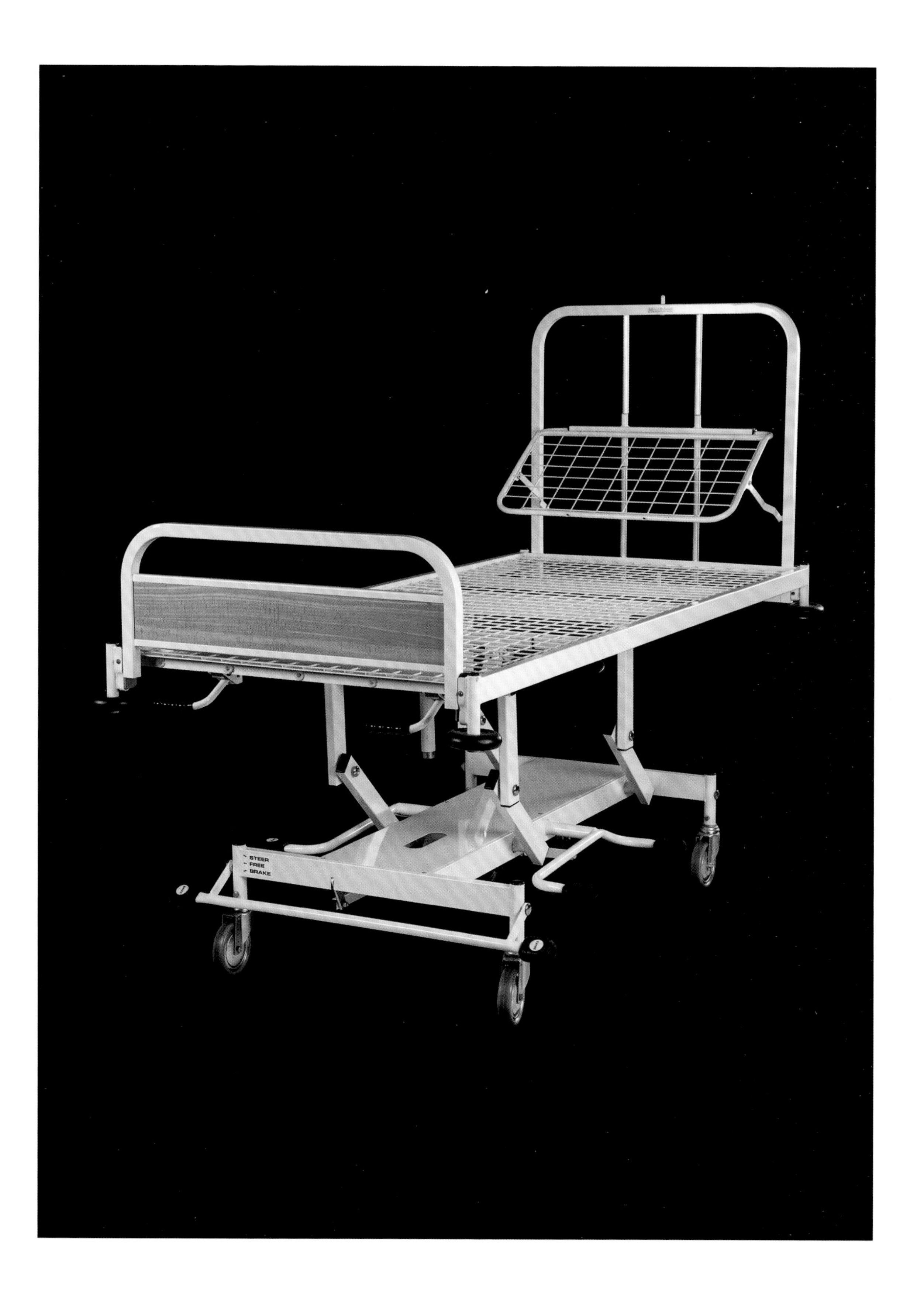
STEER
FREE
BRAKE

OFFICE OF THE METROPOLITAN COMMISSIONERS IN LUNACY'S WRITING DESK

Officers from the Metropolitan Commissioners in Lunacy had a lot of paperwork following the Madhouses Act of 1774.

Writing desks like this were designed to be portable and often stored the owner's writing equipment. This particular example came from the Office of the Metropolitan Commissioners in Lunacy and contains relevant Acts of Parliament and oaths used for giving evidence. A testament to the treatment and status of mental illness in the 1800s, the documents contained within this desk use language deemed unacceptable today. Terms such as "rogues, vagabonds and other idle and disorderly persons" are littered throughout the paperwork. If considered to have a mental illness, individuals could be "kept safely locked up in some secure place" and if necessary "be there chained". The Commission was formed by the 1828 Madhouses Act to inspect private asylums in Middlesex and the London metropolitan area, long after these institutions were established.

By the mid-eighteenth century it was common for those considered insane to be confined to their family homes or placed in an asylum or "madhouse". At the time madhouses were little more than private homes, whose owners were paid to detain their residents. Often referred to as the "trade in lunacy", private asylums proliferated throughout the UK, set up by families or individuals who saw their commercial value. While some professed medical training and a genuine concern for those affected, many were run with little or no medical involvement. These ungoverned institutions led to thousands of people being detained in inhumane conditions. Various scandals in the mid-1750s led to the first piece of official British legislation concerning mental health: the Madhouses Act of 1774. This law empowered an inspecting body to carry out visitations, required all private madhouses to have a licence and, perhaps most importantly, necessitated a letter from a medical practitioner for lawful confinement.

Although this legislation made it possible to keep track of who was confined and where, it was insufficient in preventing the abuses rife within asylums. The formation of the Metropolitan Commissioners in Lunacy in 1828 gave commissioners the power to set minimum acceptable standards for asylums. Consisting of doctors and lawyers, the organization took it upon themselves to improve and standardize medical treatment of patients. Every year the Commission had to submit reports of its findings to the Home Secretary. Writing desks like these played an important role in the process of reporting and raising standards in the Victorian asylum.

Opposite: Writing desk from the Office of the Metropolitan Commissioners in Lunacy, England, 1828–1914.

Left: Colney Hatch Lunatic Asylum, Southgate, nineteenth century.

THE SMITH-CLARKE IRON LUNG

For those unable to breathe while their chest muscles were paralysed by polio, an iron lung allowed a hope for recovery.

It is hard for us today to look at an iron lung without catching our breath. To be trapped and powerless, with our breathing outside our own control, is a fear that haunts our dreams. But for people infected by polio and unable to breathe on their own, the iron lung machine was literally a life saver. The virus poliomyelitis, commonly known as polio or infantile paralysis, was primarily a disease of young children and was most active from the late-nineteenth to the mid-twentieth centuries, at which point an effective vaccine was discovered. Polio can kill the body's motor neurons, meaning that control of the muscles is lost. Often this is temporary but sometimes it can be permanent. When the paralyzed muscles included those of the respiratory system, death by suffocation could occur, and so inventing artificial ways of keeping the patient alive while the muscles recovered became a priority for doctors and scientists.

The theory behind an iron lung is simple. The chest and torso are contained in an airtight space with the head outside. Using a pump, negative air pressure is alternately created and removed, compelling the chest to compress and relax, forcing air in and out of the body. The first effective model, built by Philip Drinker in 1927, was challenging for both patients and nurses. Replacing the Drinker model was the Both iron lung whose coffin-like appearance did little to reassure nervous patients. Nurses found it difficult to care for their patients as there were only two small access ports and the patient's head had to be forced through a tight rubber collar, causing yet more distress. The race was on to design an improved version.

Captain G.T. Smith-Clarke had recently retired as chief engineer at Alvis Motor Company when he joined the board of Warwickshire Hospital Management Committee in 1951. His version of the iron lung was nicknamed "the alligator", since the hinged chamber divided horizontally, resembling an alligator's jaw. Smith-Clarke's motor car connections came in handy; he borrowed the design of the quick-release petrol caps used on racing cars as covers for the nursing ports. He was also able to factor comfort into the design, including improved head rests and an internal stretcher system for the patient to lie on. Thankfully very few iron lungs are in use today, and many only exist within museums. The Science Museum boasts the largest iron lung collection in the world.

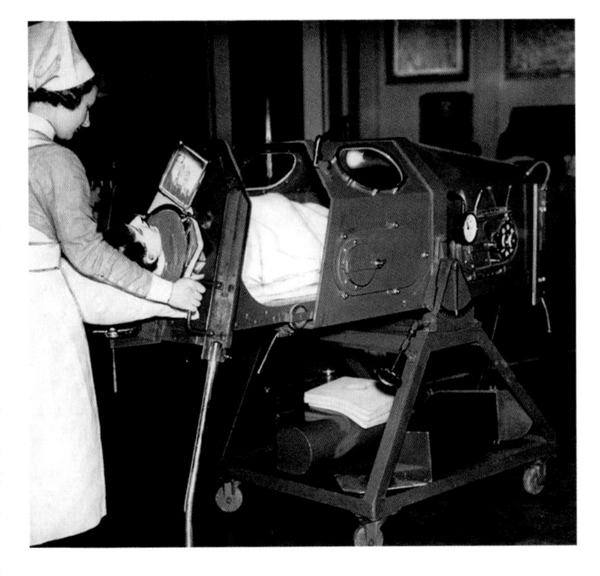

Above: A polio patient being placed in an iron lung, 1938.

Opposite: "Smith-Clarke Senior" adult cabinet iron lung, 1953.

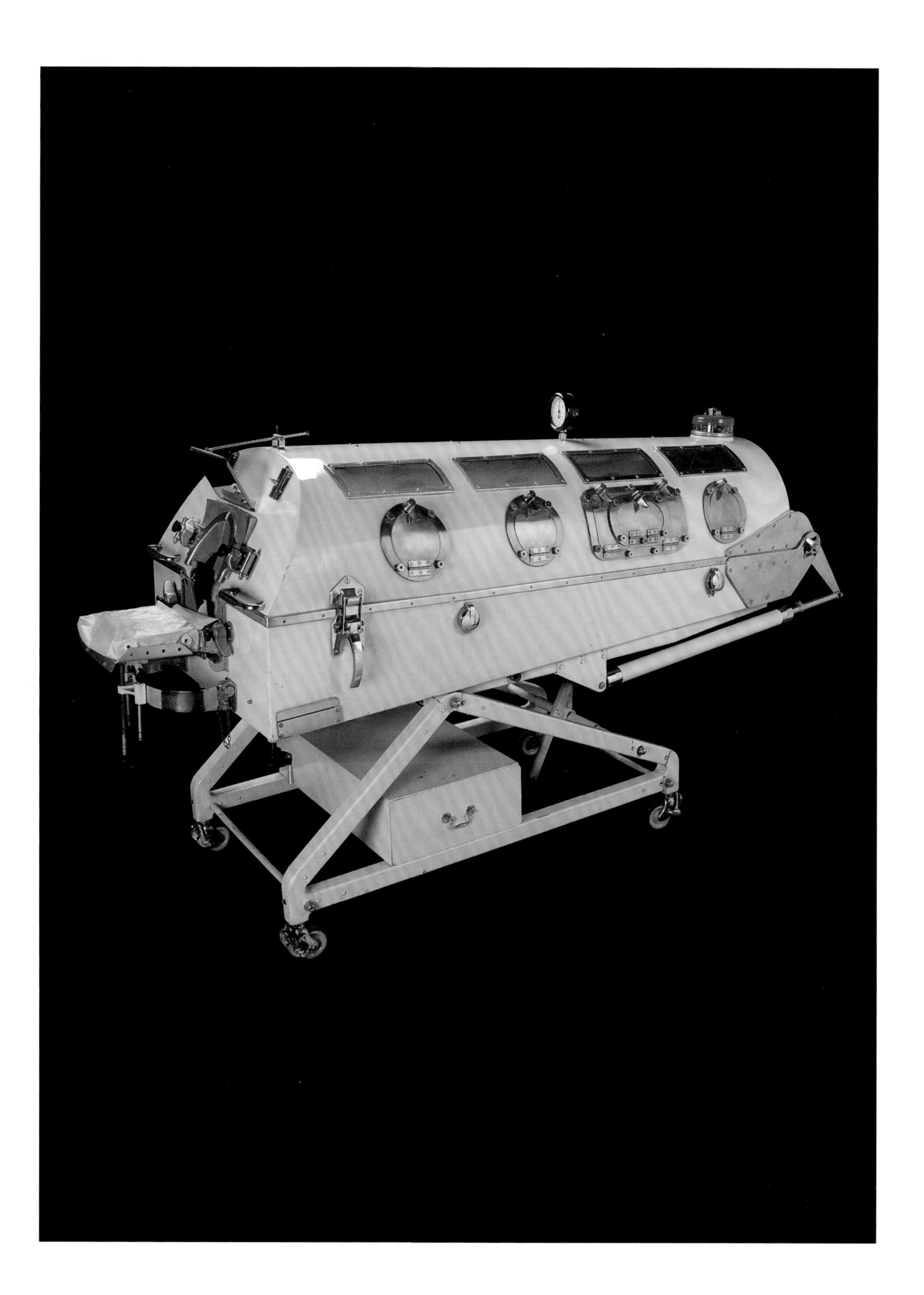

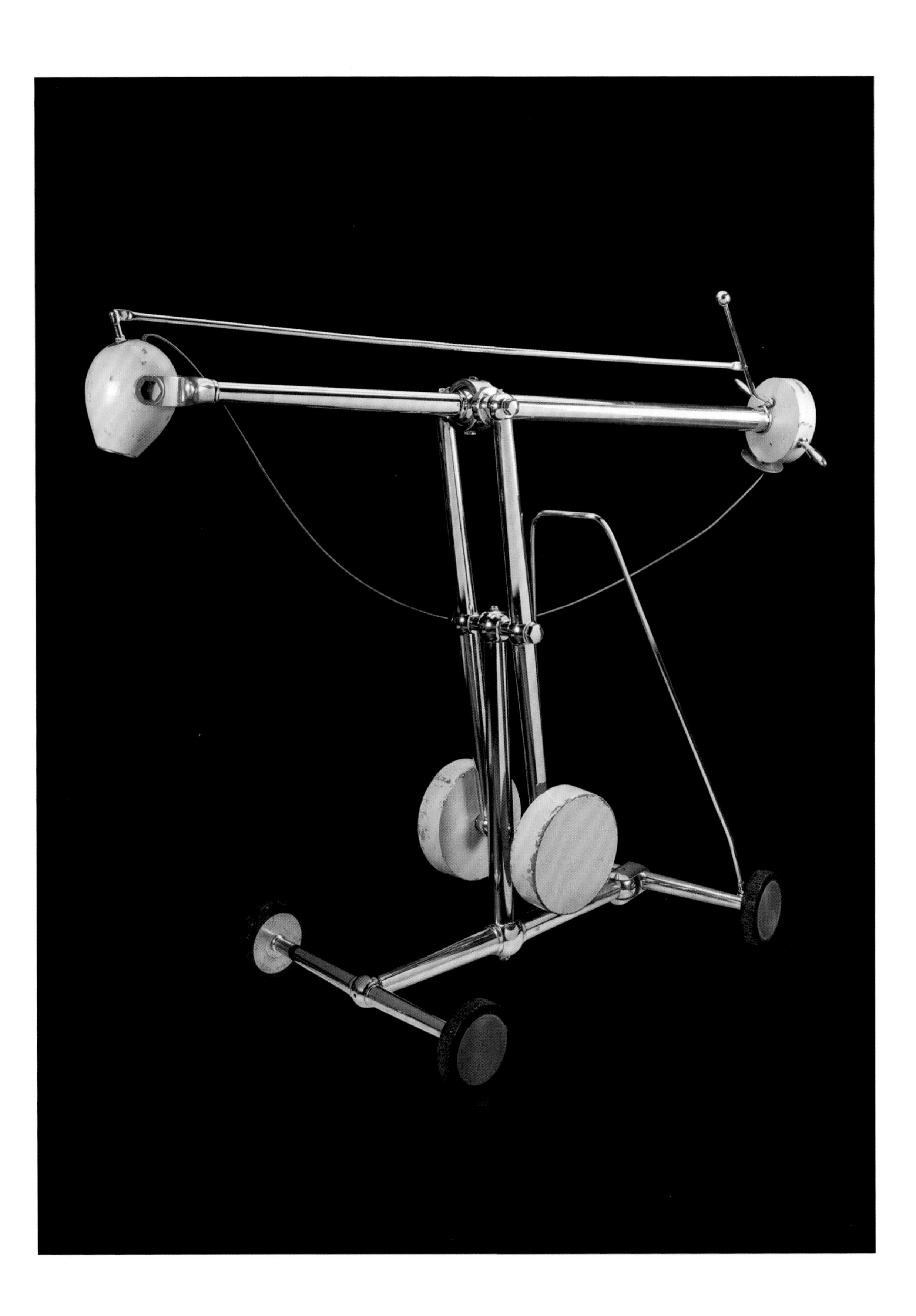

THE RADIUM "BOMB"

Ernest Rock Carling invented the "Radium Bomb" to bombard cancer of the head and neck with radiation.

Known by its makers as a radium "bomb", this makeshift-looking machine is the fourth attempt of a team of surgeons, engineers and physicists at making a teletherapy device to treat head and neck cancers. Teletherapy is the use of radiation from outside the body to treat cancer. Perhaps the bombardment of cancer with radiation gave the device its name. In this "bomb" radium was placed in the egg-shaped, lead-lined head using a shutter operated by a bicycle brake cable. The lead lining limited harmful radiation affecting healthy tissue.

Since the turn of the twentieth century, cancer has been a public health concern. As people began living longer in better housing, and with increasing control over infectious disease, soaring numbers were diagnosed with cancer. Funding into treatment options became available to research answers to this problem. Surgeon Ernest Rock Carling became interested in the therapeutic uses of radium in 1920 after visits to hospitals in continental Europe. Along with his son Francis and colleagues Stanford Cade, a surgeon, and Frank Allchin, a physician, Carling developed the first radium "bombs", experimenting with materials, doses of radium and how best to position the patient. The Science Museum cares for the second, third and fourth models. It is likely that the first model was used for its parts to build the latter versions.

Radium was expensive – a single gram (0.04 oz) could cost up to £14,000 in the 1920s. British governmental and scientific bodies were set up to ensure that the expensive and scarce resource was being used effectively. The Royal Commission, now known as The King's Fund, had loaned the Westminster team a four-gram Belgian radium bomb in the late 1920s. After 15 months the bomb was not yielding results, so the radium was split between four hospitals, resulting in this one-gram Westminster version. Physicists Henry T. Flint and Leonard Grimmett, with the existing team acting as advisers, developed this fourth version. They reduced its size and weight, making it easier to handle, and for first time the radiation source was being taken to the patient rather than the patient to the source.

Variations and modifications of the radium bomb remained in use across Europe until 1953, by which time new sources of radiation had become available, such as cobalt-60, which were safer, 300 times stronger, easier to handle and cheaper. Many of the team behind this early innovation would make cancer treatment their life's work.

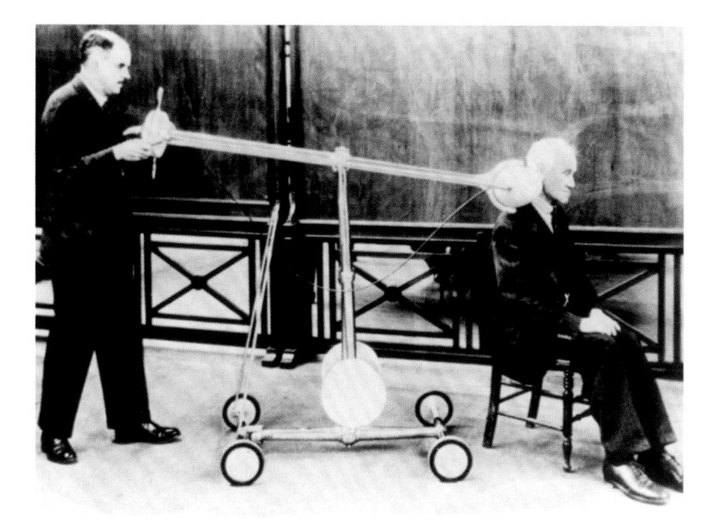

Opposite: The fourth Westminster radium "bomb", *c.*1930.

Left: Man with neck cancer receiving radiotherapy treatment from a Flint radium "bomb", designed in 1934.

THE FIRST BICYCLE AMBULANCE

Former BMX champion Tom Lynch knew he could reach medical emergencies faster on his bike than in an ambulance.

If you were to dial 999 in a medical emergency, you would probably not be expecting a bicycle to arrive. Yet ambulance bicycles are becoming a familiar sight in city centres, at airport terminals and at outdoor events. Pedalling paramedics have two distinct advantages: the ability to cut through heavy traffic and the ability to access areas that motorized vehicles cannot reach. London's Ambulance Cycle Response Unit (CRU) was the first of its kind in the world. The brainchild of Tom Lynch MBE – an Emergency Medical Technician and former British, European and World Team BMX racing champion – the service has grown from a one-man operation in 2000 to a team of more than 100 trained responders.

Lynch joined the London Ambulance Service after retiring from BMX competing in the early 1990s. But as someone used to travelling under his own steam, being confined to a vehicle felt stifling. Even more frustrating was the near-constant London traffic, which inevitably delayed patients' access to life-saving care. Convinced he could reach people faster by bike, Tom managed to persuade his bosses to agree to a trial and the CRU was born.

This bike – a customized Cannondale – was the prototype. Friends helped him to design specialist clothing and branding to ensure both frame and rider stood out on the road. The pannier was stocked with the lightest medical equipment available, the most vital of which proved to be the portable defibrillator, developed by British inventor and cardiologist Professor Frank Pantridge CBE. At first there were concerns about how patients would react to a bicycle when they were expecting an ambulance, but these fears were quickly laid to rest. As Lynch put it: "If you're having a heart attack, you don't care who arrives." Indeed, one of the scheme's major achievements has been reducing the time taken to reach cardiac arrest victims, significantly improving survival rates in the areas it serves. At the other end of the spectrum, cycle responders are well placed to advise on self-care and other pathways that avoid unnecessary transfers to hospital – what Lynch refers to as "community ambulancing". Today, cycle responders attend about 16,000 calls a year, half of which do not require hospital treatment. And, because they are out on patrol, their average response time is six minutes or less.

Above: Tom Lynch on a BMX ambulance bicycle, 2000.

Opposite: Bicycle used in the initial trial and development of the London Ambulance Cycle Response Unit, customized and used by Tom Lynch MBE, *c.*2000.

LONDON
AMBULANCE SERVICE
TOM LYNCH

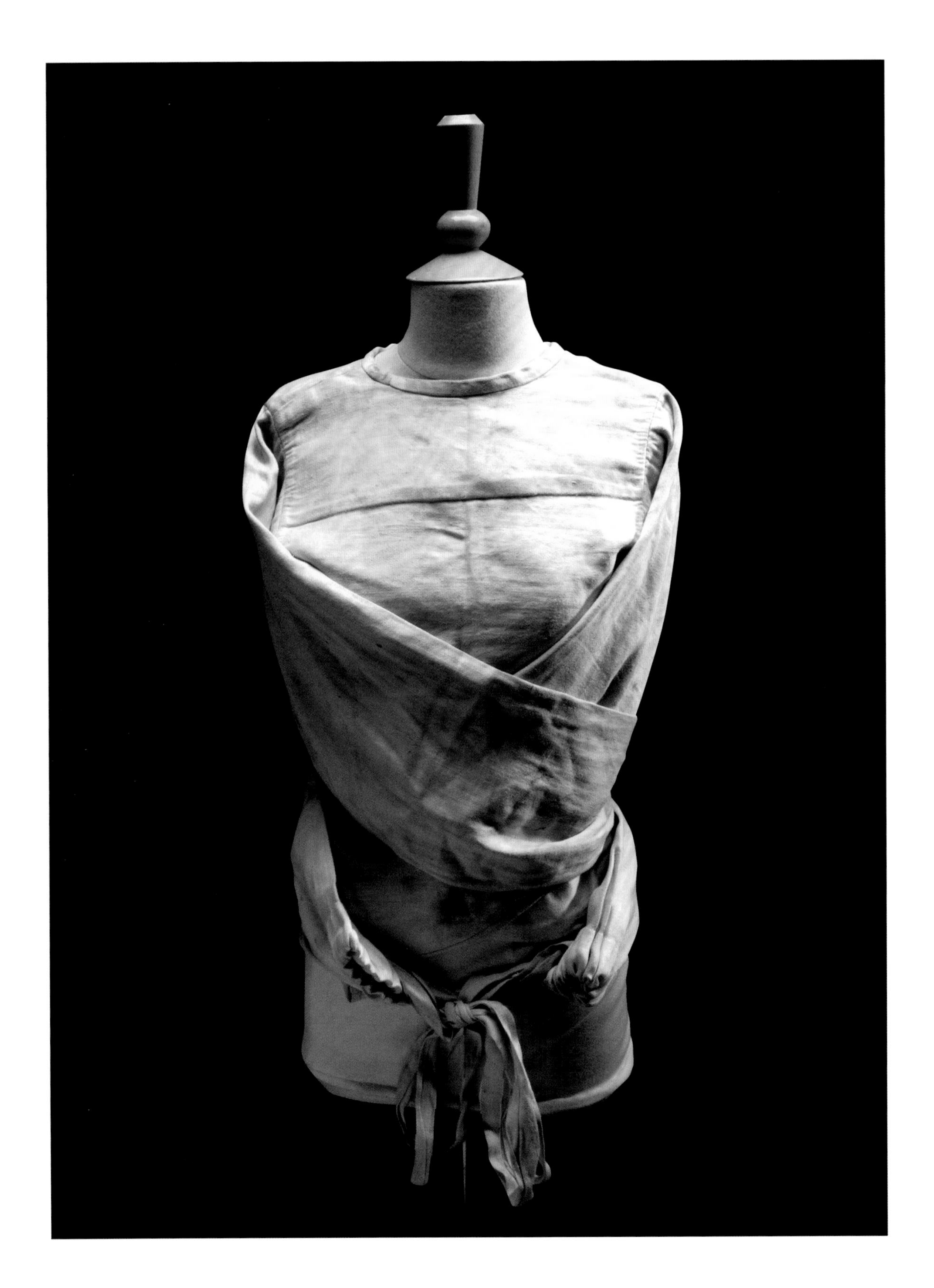

CANVAS STRAITJACKET

Restraining people in lunatic asylums using straitjackets was common practice in Victorian Britain.

"The sleeves are made tight, and so long as to cover the ends of the fingers, and are there drawn close with a string [...] by tying the sleeve-strings fast round the waist, he has no power of his hands." Referred to as a "strait waistcoat", the straitjacket was described by Irish physician David Macbride as early as 1772. At this time ideas about mental health remained abstract and theoretical, and as medical historian Roy Porter noted: "[I]t had long been assumed that the mad were like wild beasts, requiring brutal taming."

By the early nineteenth century mechanical restraints were seen as justifiable methods of controlling violent or disruptive patients in British asylums. Though definitions varied, "mechanical restraint" generally referred to any physical restraint of the body, often involving the use of straitjackets, chains, straps and coercion chairs. Garments like this were used to restrict movement and were seen as a means of pacifying unruly patients. At the height of their use, straitjackets were considered more humane than older methods of restraint involving ropes and chains. In 1814 a parliamentary committee heard that one patient, James Norris, had been restrained for nearly a decade at Bethlem Royal Hospital. It was reported that multiple apparatus had been used to confine Norris and "a stout iron ring was riveted round his neck".

It was thought that garments such as the straitjacket caused little or no harm to their wearers and afforded some freedom of movement, in contrast to patients confined to a chair or bed by straps. Although considered humane by some, straitjackets were frequently misused. As the number of patients admitted to asylums rose, there was rarely sufficient staff to provide the necessary levels of care. With attendants often lacking any training in mental health care, many resorted unnecessarily to straitjackets to maintain order and control.

The late eighteenth and early nineteenth centuries saw an emergence of new attitudes towards the management and treatment of people diagnosed as mentally ill. Robert Gardiner Hill and John Conolly independently introduced non-restraint protocols to asylums in Lincoln and Middlesex in the 1830s; these relinquished all forms of restraint, from irons and manacles to fabric cuffs and straitjackets. This straitjacket was used as recently as the 1960s to restrain adult patients in psychiatric hospitals in the UK. The development of antipsychotic drugs and the shift towards managing mental health in the community, rather than institutions, has all but eliminated the use of restraint devices.

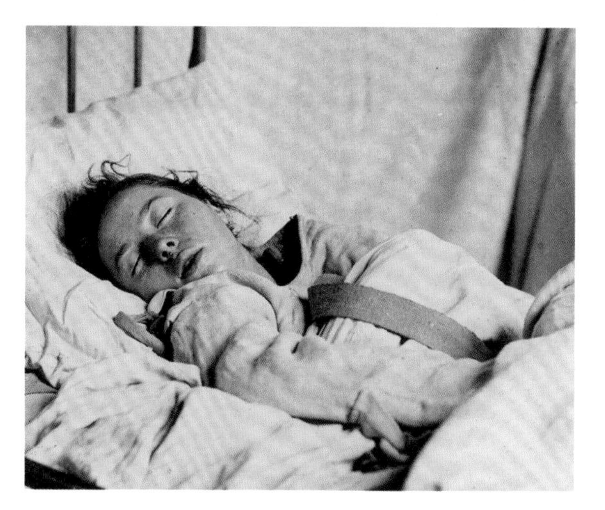

Opposite: Canvas straitjacket for restraining adult patients, made in London, England, *c.*1930–36.

Above: A female patient of Paris's Salpêtrière Hospital, in a straitjacket, 1890s.

THE FINSEN LAMP

Niels Finsen pioneered the use of light's healing properties, inventing the Finsen lamp used for treating skin conditions.

Our relationship with light – whether natural or artificial – is complex. We use light to give us the so-called sunshine vitamin (Vitamin D), which allows us to form strong bones and teeth. Too little light causes deficiency diseases such as rickets, which can lead to permanent bone deformities. Too much light, however, can burn and blister our skin, eventually causing cancers.

With the development of new technologies, light was to become a cure for disease as well as a cause of health problems. The bactericidal uses of ultraviolet light became popular in the late nineteenth and early twentieth centuries, and both natural light (heliotherapy) and artificial light (phototherapy) were used on a number of conditions, such as the scars caused by smallpox and skin diseases caused by tuberculosis, such as lupus vulgaris.

Niels Ryberg Finsen was a physician who experimented with ultraviolet light in Copenhagen in the 1890s. He deliberately burnt himself using both natural and artificial light, and tested the abilities of different materials to block ultraviolet rays. Initially Finsen worked with sunlight focused down a quartz-glass lens, but his research eventually led him to the development of this lamp. Using carbon arc lamps as a method of healing, which had been more usually used as part of industrial manufacturing, was a truly innovative concept. This large lamp created artificial ultraviolet light, delivered down telescopic tubes that contained several lenses, resulting in concentrated blue, violet and ultraviolet light, which could be shone directly on to the patient.

The Finsen lamp tube pictured was given to the Royal London Hospital by Princess Alexandra in 1900, brought from her native Denmark. Light therapy was a modern and innovative treatment that required technical equipment and a high level of knowledge. The nurses from the newly formed Light Departments of hospitals took up the challenge to not only monitor the application of the treatment but also maintain and operate the new medical technology of the Finsen lamp. Niels Finsen was awarded the Nobel Prize in 1903 for his ground-breaking invention. Light therapy is still in use today, primarily for dermatological issues, to correct irregular sleeping patterns or as a treatment for depression.

Left: One of the brass tubes from the Finsen lamp.

Opposite: Undergoing light therapy at the Royal London Hospital from James H. Sequeira's *An Elementary Treatise on the Light Treatment for Nurses*, 1905.

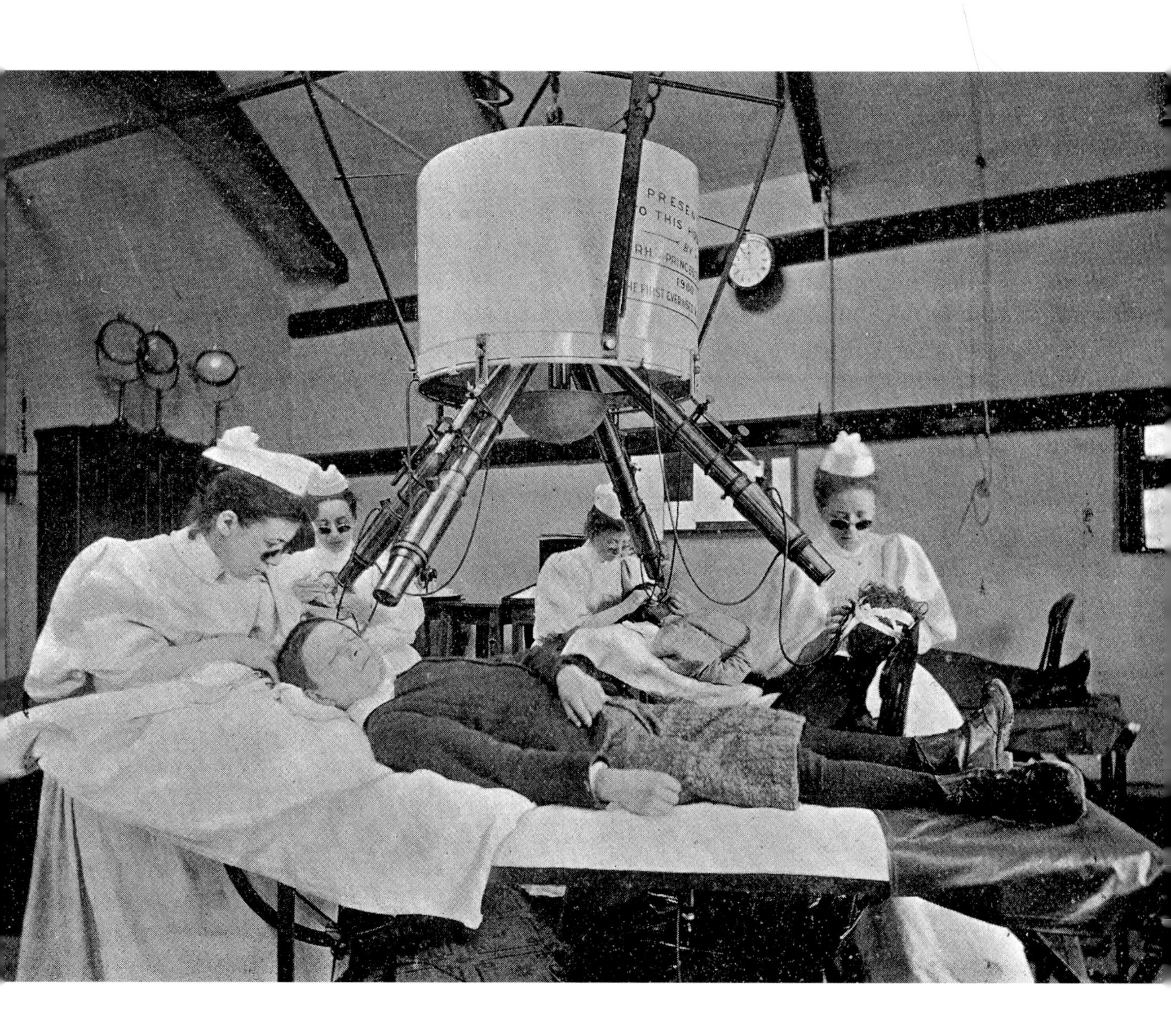

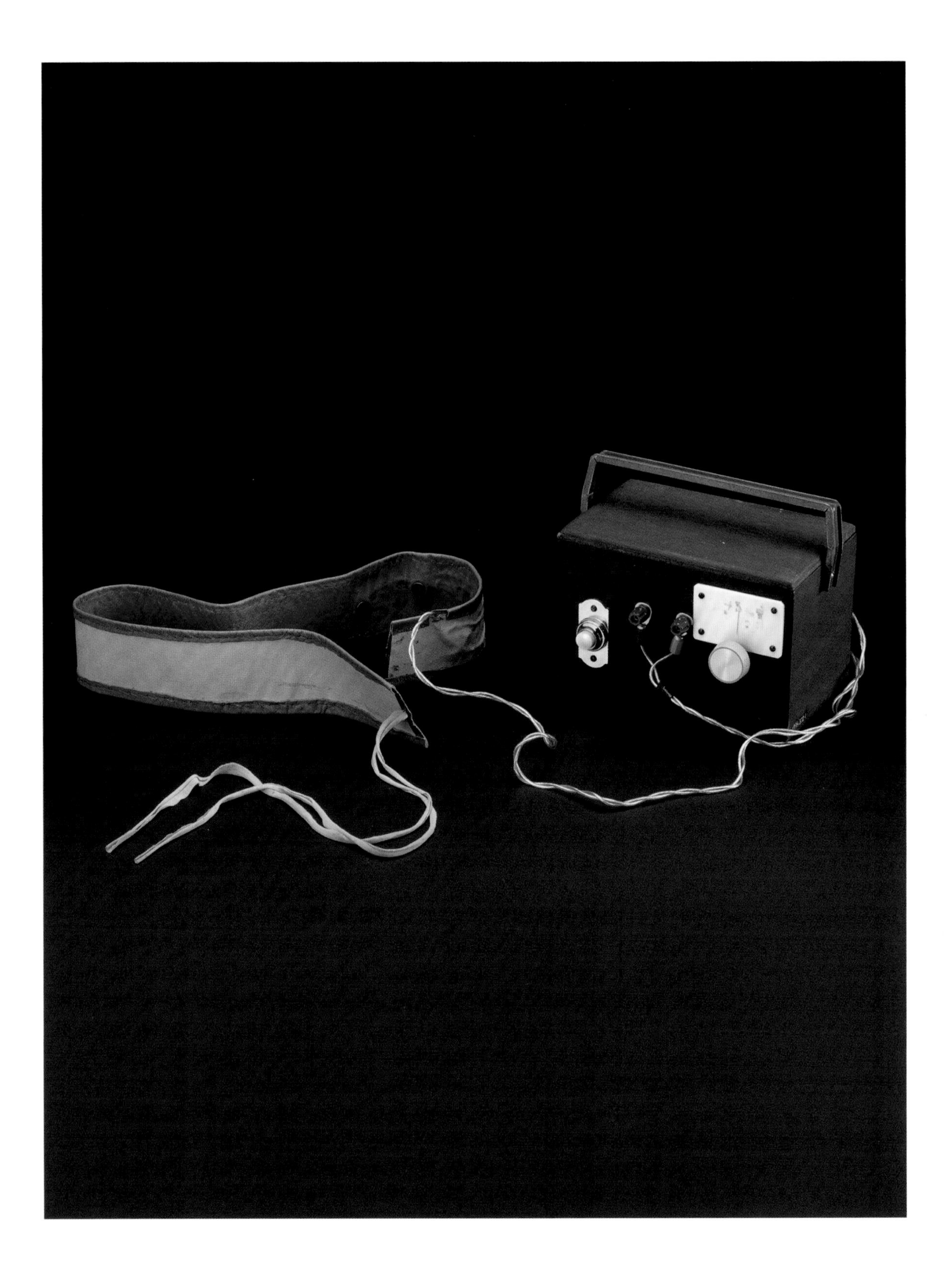

AVERSION THERAPY DEVICE

Aversion therapy involved giving short electric shocks with the intention of dissuading the patient from certain thoughts and behaviours.

Aversion therapy is a form of psychological conditioning designed to make a person avoid or give up "undesirable" habits. The theory that the mind can be conditioned to respond in a certain way, by creating negative associations with certain behaviours, is a powerful one. The concept derives primarily from the work of psychologist Ivan Pavlov, whose infamous experiments with dogs formed the basis of behaviourism, a form of psychology that dominated the mid-twentieth century.

First used in the 1930s to treat alcoholism, aversion therapy has since been used to "treat" other behaviours deemed "unacceptable", ranging from phobias and addictions through to homosexuality. As recently as 1974 homosexuality was officially considered a psychiatric illness. The concept of same-sex attraction as a medical or psychological condition was introduced at the end of the nineteenth century. This medicalization of sexuality was seen to legitimize treatments to "correct" it. This device is said to have been used in the treatment of homosexuals at a psychiatric hospital in Brighton in the mid-twentieth century. During the therapy patients were exposed to certain stimuli, such as pornographic imagery featuring people of the same sex. Simultaneously electric shocks of varying strengths and durations were administered to the person's fingers or hands. This was thought to diminish homosexual desires or even transform an individual's sexuality.

Techniques intended to "cure" those outside the so-called heterosexual norms had become commonplace within psychiatric institutions and National Health Service hospitals by the 1950s, and though sexual relationships between adult men in private were decriminalized in Britain in 1967, treatments such as aversion therapy continued to be used. Many individuals are said to have volunteered themselves for treatment, often as a result of societal pressures, while others arrived through the criminal justice system. Before 1967 gay and bisexual men convicted of homosexuality faced a maximum sentence of life in prison in the UK, and many were forced to undergo treatment as part of their sentence. Others opted to undergo treatment as an alternative to imprisonment.

Left: Aversion therapy device from St Francis Psychiatric Hospital, Sussex, England, 1950–80.

Above: Campaign for Homosexual Equality rally, 1974.

The labelling of homosexuality as a disease and a crime led to discrimination and the infringement of what we now consider basic human rights. In the 1970s gay rights activists (including psychiatrists) protested that they were in no need of being "cured". Attitudes within the psychiatric community gradually shifted and in 1973 the American Psychiatric Association removed homosexuality from its *Diagnostic and Statistical Manual of Mental Disorders.*

TEACHING DOLL TO SHOW MEDICAL TREATMENT TO CHILDREN

Treatment for children with polio often involved long bed rest, encased in plaster casts.

Serene, wide open and bright blue, this doll's eyes show that being encased in rigid plaster is nothing to be afraid of. Dolls made of china or celluloid would have been familiar to every child in the first half of the twentieth century. But instead of attending a dolls' tea party, this doll is undergoing therapy for tuberculosis of the bone or polio.

Communicating with small children in a hospital, who were probably feeling sick and scared, was a challenge. This doll – and others in the same set, all demonstrating a variety of treatments – was used to explain what was about to happen to the children who were patients at what was once known as the Lord Mayor Treloar Cripples' Hospital. Founded in 1908, its name was changed to the Lord Mayor Treloar Orthopaedic Hospital in 1951, to avoid the outdated terminology "cripples". Initially for children with surgical tuberculosis, the hospital admitted polio patients from the 1920s onwards. Both of these conditions can damage the joints, bones and muscles of growing bodies. The use of metal braces and callipers was combined with casting parts of the body in plaster, in an attempt to prevent the tightening of tendons and ligaments, which could reduce movement and affect healthy bone growth. For children whose spines were affected, a full body cast was used, often moulded on the child as they were suspended from a frame. A child might spend months in such a cast. The efficacy of this treatment was limited as, although it prevented further twisting of the bones, the forced immobility inevitably caused muscle weakness.

Toys, and perhaps especially dolls, are the receptacles of a child's imagination. But this doll is more than just a plaything; it marks a crossroads in twentieth-century child health. It was created towards the end of epidemics of polio and the dread of tuberculosis in Britain (due to the introduction of effective vaccines and penicillin) but also in the early years of modern child psychology. In the 1930s and 1940s developments in psychology included the use of play to communicate with children and showed a clearer understanding of the long-term impact of childhood trauma. The dolls' neutral expression and passive attitude would have been very different from that expressed by a real child made of flesh and bone.

Left: "Out on the terrace" – children being treated for polio and tuberculosis at the Lord Mayor Treloar Cripples' Home, Hospital and College in Alton, Hampshire, *c.*1937.

Opposite: Ceramic teaching doll to show treatment for polio, 1930–50.

THE MINIATURE HOSPITAL

The Kings Fund created this large dolls' house of a hospital to tour the country as a way of raising money for the charity.

Although populated by numerous doll-like figures, this intricate model was not made to be played with. It was commissioned in the early 1930s by a charity, then known as the King Edward's Hospital Fund for London (now The King's Fund), to inspire a public sense of pride and responsibility for hospitals. In the days before the National Health Service it also functioned as an elaborate fund-raising device. The original exhibition of the model was inaugurated by the then Prince of Wales, who was later to reign very briefly as Edward VIII. The model toured venues across the nation, transported in its own special trailer, and was seen by thousands.

The 1:16 scale model depicts the wing of an ideal modern hospital of the period. Consisting of five separate blocks that connect together for display, it can also be fully illuminated by over a hundred bulbs housed in tiny light fittings. The level of detail is breathtaking, even in parts of the hospital that can only be glimpsed through windows or at the ends of corridors. Alongside the 13,000 handmade and painted tiles, and the depiction of more than one million bricks, the specialist equipment shown throughout the model – be it X-ray machinery, operating theatre lights or kitchenware – was made in collaboration with companies manufacturing these products in the real world. Elsewhere, the framed portraits in the sycamore-wood-lined board room were completed by one of the leading miniaturist painters of the day. And, in a reflection of the high social standing of those who helped to fund and then publicize the model, several of the bedspreads in the adult ward were made from lace handkerchiefs donated by female members of the Royal Family.

In its early days, the model also had at least one moving part. By dropping a coin in a special box and pressing a button the hospital lift could be operated. The surviving paper label for this simple interactive device hints at a story familiar to us all today. Beneath the original message, a hand-written addition explains that users "Please take lift either up or down. Once only" before adding hopefully "Please. Do not use as a toy". Sadly, on its reverse are written three more simple words: "Out of order".

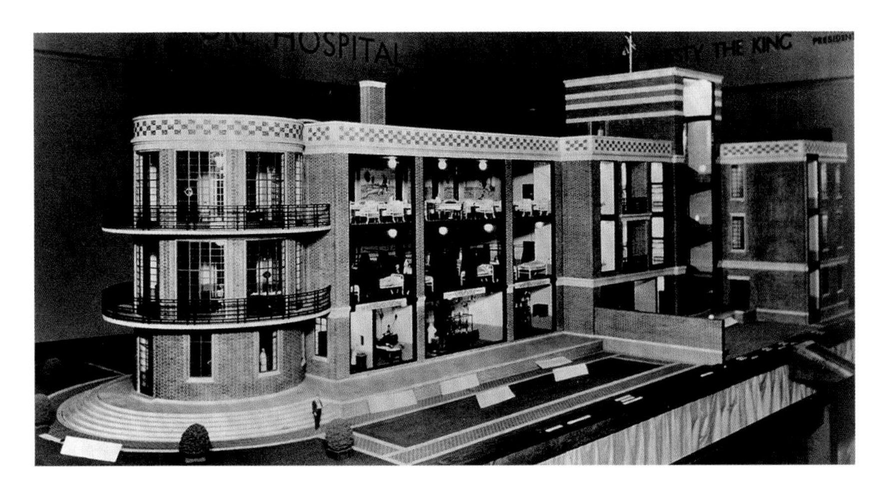

Opposite: The children's ward in the miniature hospital.

Left: The miniature hospital shown on a publicity postcard from the time of its public unveiling in 1933.

INDEX

ACKNOWLEDGEMENTS

This book is a collective effort by staff at the Science Museum and researchers actively working with the medicine collection here. It features authors from a variety of disciplines, including art historians, curators, scientists, and conservators, who are all united by a deep understanding of the medical artefacts at the Science Museum and a passion for sharing their expertise.

As editors of this book we would like to thank each of the authors for their contributions, expertise and unstinting support throughout its creation: Gemma Almond, Muriel Bailly, Katy Barret, Sarah Bond, Tim Boon, Jessica Bradford, Robert Bud, Imogen Clarke, Rupert Cole, Katie Dabin, Jack Davies, Stewart Emmens, Rebecca Kearney, Jannicke Langfeldt, Isabelle Lawrence, Emma Stirling-Middleton, Sara Stradal, Annie Thwaite and Sarah Wade.

We are indebted to Kevin Percival, Jennie Hills and Kira Zumkley, who produced the beautiful photography, and to Charlotte Grieveson and Taragh Godfrey for their editorial expertise. We are also grateful to Issy Wilkinson at Carlton Books. As ever, curators at the Science Museum benefit from sharing ideas with our friends and colleagues at Wellcome Collection, without whom this book would not have happened and we extent our thanks and gratitude to them. Finally, we would like to thank our colleagues working behind the scenes for their ongoing care and dedication to our world-class collections.

CREDITS